Lecture Notes in Economics and Mathematical Systems

Operations Research, Computer Science, Social Science

Edited by M. Beckmann, Providence, G. Goos, Karlsruhe, and H. P. Künzi, Zürich

87

G. F. Newell

Approximate Stochastic Behavior of n-Server Service Systems with Large n

Springer-Verlag
Berlin · Heidelberg · New York 1973

Dr. Gordon F. Newell
University of California
Institute of Transportation and
Traffic Engineering
109 McLaughlin Hall
Berkeley, CA 94720/USA

AMS Subject Classifications (1970): 60 K 25, 60 K 30, 62 M 10, 90 B 99, 94 A 20

ISBN 3-540-06366-8 Springer-Verlag Berlin · Heidelberg · New York
ISBN 0-387-06366-8 Springer-Verlag New York · Heidelberg · Berlin

PREFACE

For many stochastic service systems, service capacities large enough to serve some given customer demand is achieved simply by providing multiple servers of low capacity; for example, toll plazas have many toll collectors, banks have many tellers, bus lines have many buses, etc. If queueing exists and the typical queue size is large compared with the number n of servers, all servers are kept busy most of the time and the service behaves like some "effective" single server with mean service time $1/n$ times that of an actual server. The behavior of the queueing system can be described, at least approximately, by use of known results from the much studied single-channel queueing system. For $n \gg 1$, however, (we are thinking particularly of cases in which $n \gtrsim 10$), the system may be rather congested and quite sensitive to variations in demand even when the average queue is small compared with n. The behavior of such a system will, generally, differ quite significantly from any "equivalent" single-server system.

The following study deals with what, in the customary classification of queueing systems, is called the $G/G/n$ system; n servers in parallel with independent service times serving a fairly general type of customer arrival process. The arrival rate of customers may be time-dependent; particular attention is given to time dependence typical of a "rush hour" in which the arrival rate has a single maximum possibly exceeding the capacity of the service.

The methods of analysis exploit a postulate that $n \gg 1$, that all relevant counts of customers are made on a scale which is also large compared with 1 (typically of order n), and that stochastic fluctuations in arrival counts of order n are of order $n^{1/2}$. Graphs of cumulative counts of customer arrivals and available servers are used to represent the evolution of stochastic realizations of the system. A combination of graphical and analytic methods are then used to estimate queue distributions for various typical types of behaviors.

If the traffic intensity (arrival rate/service rate), $\rho(t)$, increases toward a maximum, the system behaves like an ∞-channel system until $1 - \rho(t)$ is of order

$n^{-1/2}$. When $\rho(t)$ come sufficiently close to 1 or exceeds 1 so that queueing becomes virtually certain, the system then behaves essentially like some effective single-channel service system. While $\rho(t)$ is between these two extremes, when it is uncertain whether or not a queue exists, several different types of behavior can exist depending upon how long the system stays in this transition state, and how $\rho(t)$ behaves during this time.

Chapter I describes the graphical representations and the postulates upon which the approximations will be based. Chapter II deals mostly with situations in which the above transition lasts for a time which is small compared with an expected service time. This is particularly appropriate for systems such as bus routes in which the service time (trip time) is comparable with the duration of the rush hour. In this case the distribution of the number of customers in the system (server plus the queue) remains approximately normally distributed at all times with appropriate time-dependent means and variances.

Chapter III treats the cases in which the transition lasts for a time large compared with the service time. The behavior through the transition can, for the most part, be described by "diffusion approximations." This is described very qualitatively because it is quite similar to previously analysed behaviors of a single-channel service system. Chapter IV deals with extremely slow changes in $\rho(t)$, equilibrium distributions. Approximate distributions are compared with exact distributions for the $M/M/n$ and $M/D/n$ systems (Poisson arrivals and either exponentially distributed or deterministic service times).

Despite the length of this report, it contains very few detailed results. It describes mostly a classification of types of behaviors and the qualitative properties of these types, and methods which could be used to obtain more quantitative results. It would have been desirable to add a few "case studies" to test the accuracy, but, as yet, none have been made. The only numerical comparisons are with the known equilibrium distributions mentioned above. It would appear, however, that for $n \gtrsim 10$, any person skilled in the art of making rough calculations should be able to estimate average lengths of queues, delays, etc., to within about 10% in most situations he is likely to encounter; this is the level of accuracy needed in most design applications.

Most of the preliminary studies upon which the following is based were made over a period of several years while I was trying to teach transportation engineering students some uses of mathematical methods of analysis. The style clearly reflects the consequences of attempting to show both theoreticians and "hard-nosed" engineers how to find useful answers to real problems. I am indebted to my colleagues and students of diverse backgrounds for forcing me to address my remarks to all of them simultaneously.

This work was not directly supported by any research grants, however, student support subsidiary to this work has been financed by the National Science Foundation under Grants GP 9323 and GP 24617. Miscellaneous expenses have also been financed in part by a grant from General Motors Corporation. Most of the following was composed while I was on sabbatical leave January-July 1972 in residence at Union College, Schenectady, N. Y. I am indebted to Union College, particularly to Professor Gilbert Harlow and the Department of Civil Engineering, for providing space for me and a couple of students, and a pleasant atmosphere in which to work.

The final editing was done at Berkeley; the typing was done by Phyllis De Fabio in the Institute of Transportation and Traffic Engineering.

CONTENTS

$$\text{CHAPTER I - GENERAL FORMULATION}$$

1. Introduction

With but a few exceptions, the existing literature on multiple-channel service systems has not been very useful in the analysis of typical practical problems for at least three reasons. First, despite the existence of a few thousand papers and a dozen or more books on queueing theory and related subjects, the range of problems that have been studied is rather narrow. Secondly, in the mathematical formulations, people usually make many idealizations (Poisson arrivals, stationary behavior, independent service times, etc.) which are of such doubtful validity that it is difficult to judge the accuracy with which any conclusions can be applied to practical problems. Thirdly, particularly for non-stationary behavior, the "solutions" describing the evolution of even the simplest systems are usually so complicated as to be useless for computation.

The deficiencies in applications are not so much a result of a failure to apply known techniques, as the lack of sufficiently powerful techniques to apply. Yet there are so many physical systems involving multiple servers, and such a great need for practical tools of analysis of real systems, that the problems should not be avoided just because they cannot be treated elegantly or accurately. Even a crude analysis of a realistic problem is better than an accurate analysis of a hypothetical problem.

We will be concerned here primarily with certain qualitative properties of what, in the customary language of queueing theory, would be called a multiple-channel queueing system for which the arrival process may have a time-dependent arrival rate (in addition to being stochastic), and each service channel may serve either single customers or customers in bulk. It will be assumed, however, that the number of servers (channels), n , is large compared with 1 ; that any relevant counts of customers or servers (in queue, in use, etc.) are large compared with 1 and can be approximated by a continuous variable; and that stochastic fluctuations in counts are small in some sense, though not necessarily negligible (small compared with n but large compared with 1 , for example of order $n^{1/2}$) .

Although we will use the language of queueing theory (customers, servers, queues, etc.) in describing the behavior of the system, one should perhaps think also in more abstract terms. The basic feature common to all systems to be considered here is that there are two types of objects, objects of the same type being identical. One type will be called "customer" and the other type "server," but it will not always be clear as to which label should go with which type. Each type of object satisfies a "conservation principle," i.e., no object disintegrates or changes type; if it leaves one place, it must appear somewhere else. All objects of the same type move through a specific sequence of stages (for example, customers may join a queue, enter service, and leave service), and, in at least one of these stages, the two types of objects interact, for example, one type of object cannot move unless one or more of the other type moves simultaneously. The purpose of the analysis is to determine, for any given rules of interaction, how many objects of any type are in each stage, how long they stay there, etc., or, more specifically, the probability distributions for these quantities.

In the conventional n-channel queueing system, the customers enter the system from an external source according to a given (stochastic) pattern. Each customer immediately joins a queue, then (immediately or later) enters the service, and, finally, at some later time exits from the system. The other "objects" are the n servers which enter service (becomes busy), then leave service (become idle), and then, eventually, reenter service again. The main difference between the two types of objects is that for customers there is a (perhaps hypothetical) reservoir of infinite capacity from which customers enter the system and to which they are returned. For the servers, this reservoir has zero capacity; a server which completes service immediately joins the "queue" of idle servers. It may be convenient to think of the motion of the system as represented schematically in Fig. I.1.

If the servers serve only one customer at a time, the rule of interaction between customers and servers is that each time a server enters service, it must carry a customer with it (and vice-versa). This can happen only if it is possible to pair a customer in queue with an idle server. These two remain paired for a specified time (service time), after which the server becomes idle and the customer exits. Of course, the rules of behavior also require that the number of customers in queue and

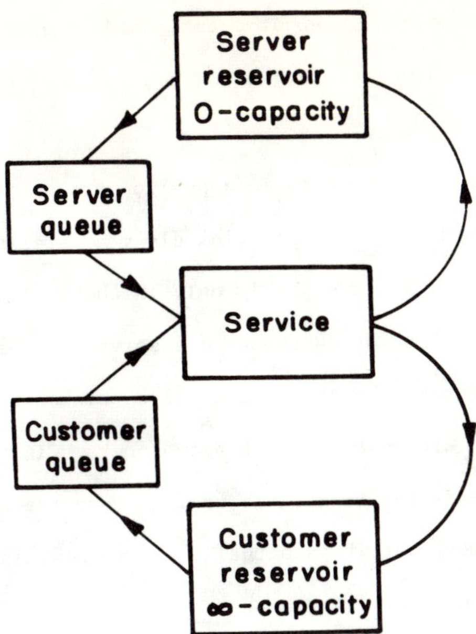

Fig. I.1 - Schematic representation
of the flow of customers and servers
in an n-server service system.

the number of idle servers shall be non-negative at all times, and the number of

customers in service be equal to the number of servers in service.

In analyzing an n-server system, we would like to imagine that n is arbi-

trary large, preferably 10^2 or 10^3 or more, but the physical systems that we par-

ticularly have in mind are the ones in which n is likely to be in the range of

only about 10 or 20. Some typical examples are the following:

1. Shoppers (customers) queue at the check-out counter of a large gro-

   cery store with many cashiers, or bank customers queue for service

   from any of many bank tellers (servers).

2. Cars (customers) arrive at a toll plaza of a highway bridge at which

   there may be about 20 toll collectors (servers).

3. Airplanes (customers) arrive at an airport terminal area and must

   queue on the taxiways for gate positions (servers) to unload

   passengers.

4. Telephone calls (customers) are held (for a short time, at least)

   for any of many possible telephone lines (servers).

5. Passengers (customers) queue at a taxi-loading zone for any of
   finitely many taxis (servers) which are assigned to serve only
   one source of customers.

6. Cars (customers) wish to park in a public garage for short time
   parking, the garage having only finitely many stalls (servers).

Of the above examples, it is probably only in the taxi service that it seems
intuitively appealing to think of customers and servers as objects with perhaps in-
terchangeable labels, because the taxi driver also thinks of himself as queueing for
the customers when there are more taxis than customers. In this case, the customer
will physically exit the system (leave the taxi) before the taxi returns to the load
point for another customer, but it is actually the round trip time from the load
point that is the relevant "service time."

A bulk service n-server service system differs from the above in that a server
is capable of carrying into service a finite number of customers greater than 1 at a
time (there might even be some systems in which a customer takes several servers
into service at a time). There is a wide variety of possible rules, however, for
deciding under what conditions an idle server will enter service. The following
analysis is, to some extent, motivated by the most obvious physical examples: almost any type of public transportation system operating on a fixed route with n
vehicles, a bus route, airline shuttle, airport limousine, train, or a bank of elevators. In this context, various rules for use of idle servers would be described
as dispatching policies. Some examples of these are:

1. A vehicle is dispatched at a scheduled time if a vehicle is
   available, otherwise as soon thereafter as possible. If no customers request service, the vehicle might be dispatched empty, or
   it may wait until it has a customer to serve.

2. A vehicle capable of serving a batch of b customers at a time
   may be dispatched only when full.

3. Any idle vehicle will be dispatched immediately if there is a
   queue of customers, taking with it all customers in queue up to
   its capacity.

4. A vehicle is dispatched according to some optimal policy which
   maximizes some suitable objective function, such as negative of
   the total passenger wait (minimize delay). Such a policy is likely
   to dispatch a vehicle only when the batch of customers exceeds some
   minimum value dependent upon the number of idle servers and expected
   future arrival rates.

There are many other types of queueing systems in which there is an interaction
between two types of objects. For example, in the "machine repairmen" problem[13]
one has a fixed number of machines and a fixed number of repairmen. Either machines
or repairmen may be busy or idle; machines may wait to be repaired or repairmen wait
for machine breakdown. In such situations, the rules of interaction are such that
there is no way of deciding which type of object should be labeled customer and
which server. We will limit the discussion here, however, to some of the typical
types of problems that arise in the analysis of some of the physical systems listed
above.

It usually costs more to build a large service than a small one, and most fa-
cilities are used to capacity only during a "rush hour." Most of the important design problems therefore deal, directly or indirectly, with the question of how large
an n is necessary to achieve some suitable compromise between rush hour delays to
customers and construction of a facility that will be idle most of the time (off
rush hour delays to servers). As a preliminary step to this comparison, one should
use "queueing theory" mainly to evaluate customer delay during the rush hour when
the arrival rate of customers approaches or exceeds the capacity of the service.
During such times, if $n \gg 1$, it will also be true that the number of busy servers
and the number of customers in the system (in service or in queue) will be large
compared with 1. We will exploit this by disregarding the fact that customers must
be integer valued. Both customers and servers will be treated as a continuous fluid
with either deterministic or stochastic properties.

One can immediately recognize that there are at least three characteristic
"time constants" associated with the rush hour behavior of systems of the type described above; the average time interval between arrivals of customers (at the peak

of the rush, for example), the average service time of a server, and the duration of the rush hour. The assumption that many servers are kept busy implies that the interarrival time is small compared with the service time. In most applications, the duration of the rush hour will be larger yet, typically in the range of 1/4 to 2 hours as implied by the designation as "rush hour" (although there are certain similar problems in production and inventory theory where the rush may be a seasonal one), while the average service time might be a few seconds for toll collectors on a highway; a few minutes for a cashier, a bank teller, or a telephone call; or a significant part of the rush hour for many public transportation systems.

In the following sections of Chapter I, we will be concerned with fairly general methods of formulation of problems of the type described above. In section 2, we will show some methods of representing the evolution of the system graphically, first for the special case in which the service times S_j for the jth service is the same for all j , $S_j = s$, and then for the more general case of different S_j . In section 3, we will discuss some of the postulates that one might make about the stochastic properties of the customer arrivals and service times and some methods of approximations that will be used, first to estimate properties of the system when stochastic effects can be neglected completely, secondly to estimate some order of magnitudes of stochastic properties, and finally, where necessary, to obtain quantitative estimates of stochastic effects.

In Chapters II-IV, we will consider in some detail a narrower class of systems in which each server serves only one customer at a time (or equivalent systems of fixed batch size). Chapter II will deal mostly with approximations and qualitative properties of the time-dependent behavior of such systems when either there is no queue of customers or a queue does exist but the expected service time is comparable with the duration of the rush hour. Chapter III will deal with time-dependent queueing when the expected service time is short compared with the duration of the rush hour and the arrival rate of customers is slowly varying on a time scale of the order of the service time. Chapter IV will consider in more detail some properties, particularly equilibrium distributions, for special systems with either deterministic ($S_j = s$) service times or exponentially distributed service times.

2. Graphical Representations

Many of the clues to an approximate analysis of the n-server service system
with stochastic properties can be seen from graphical representations of the behavior of a system with specified arrival pattern and service times. Although we
will think of the arrival pattern and service times as "deterministic" for the
present, they will later be interpreted as typical realizations of stochastic
behavior.

We first draw a graph, as in Fig. I.2a, of the (given) function

$$A_c(t) = \text{cumulative number of customers to arrive at the system by time } t .$$

This is a step function which increases by a unit jump at each arrival time. Since
we will usually be drawing this on a coarse scale of counts, it will often be approximated by a smooth curve.

The graphical constructions to follow will involve additions and subtractions
in both vertical and horizontal directions in the (t , j) plane. $A_c(t)$ can be
considered as a curve in the two-dimensional (t , j) plane or a mapping, but if it
is considered a "function of t" , one should become equally accustomed to thinking
of it as a representation of the inverse functions

$$A_c^{-1}(j) = \text{arrival time of the } j^{th} \text{ customer,}$$

the same curve as $A_c(t)$ but viewed from the j-axis.

From some description of the service mechanism, we wish eventually to construct,
on the same graph a curve of

$$D_c(t) = \text{cumulative number of customers that have left the queue and entered service by time } t$$

or, equivalently, its inverse

$$D_c^{-1}(j) = \text{time at which } j \text{ customers have entered service.}$$

The curve $D_c(t)$ is also a step function, but if customers enter the service in
batches (as for buses or elevators), the steps may be some integer greater than 1.

If we start at time 0 with no customers in the system, then at any time t ,

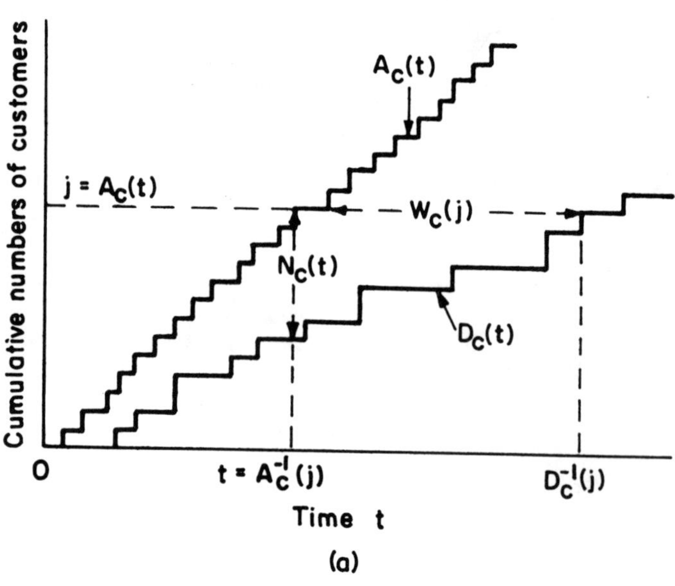

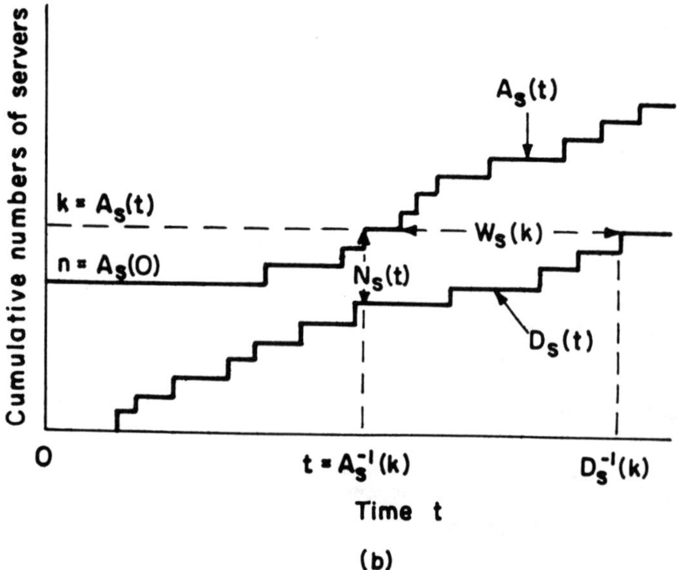

Fig. I.2 - Curves of the cumulative number of customers (a) and servers (b) to arrive by time t , $A_C(t)$ and $A_S(t)$; and the cumulative number of customers (a) and servers (b) to enter service by time t , $D_C(t)$ and $D_S(t)$. The number of customers or servers waiting to enter service at time t is $N_C(t)$ or $N_S(t)$; the wait of the jth customer or kth server is $W_C(j)$ or $W_S(k)$.

the number of customers in queue, $N_c(t)$, is the number that have arrived minus the number that have left.

$$N_c(t) = A_c(t) - D_c(t) , \qquad (2.1)$$

the vertical distance between the two curves at time t .

If the queue discipline of customers is first-in-first-out (FIFO), then the jth customer to arrive is also the jth customer to enter the service. He arrived at time $A_c^{-1}(j)$ and left the queue at time $D_c^{-1}(j)$. Thus his wait in queue is

$$W_c(j) = D_c^{-1}(j) - A_c^{-1}(j) ,$$

the horizontal distance between the two curves at height j .

We could also draw a curve to represent the times at which the jth customers left the service. If the service time of the jth customer is S_j^* , then he will leave the service at time $D_c^{-1}(j) + S_j^*$. These times, themselves, are not usually of much interest, however; they have no influence on the graphical constructions because the customer leaves the system. The service times are certainly important but it is more appropriate to identify them with the servers than with the customers because the server stays in the system.

We will also wish, eventually, to construct a curve of the function

$$A_s(t) = \text{cumulative number of servers that were available by time } t .$$

We may either draw this on the same graph as $A_c(t)$ or on a separate graph as in Fig. I.2b. This is also a step function, usually with unit steps. But if at time $t = 0$, before any customers have arrived, all servers are free, then $A_s(0) = n$ and $A_s(t)$ increases whenever a service has been completed by some server and that server has become available again. The inverse of this is

$$A_s^{-1}(k) = \text{time when the cumulative number of available servers reaches } k .$$

Finally, we also wish to consider the curve

$$D_s(t) = \text{cumulative number of servers that were used by time } t ,$$

and its inverse

$$D_s^{-1}(k) = \text{time at which } k \text{ servers have entered service.}$$

As in the case of $A_c(t)$ and $D_c(t)$, the number of idle servers at time t , the queue of servers, is

$$N_s(t) = A_s(t) - D_s(t) , \qquad (2.2)$$

the vertical distance between the curves $A_s(t)$ and $D_s(t)$ at time t . If the queue discipline among queueing servers is also FIFO, then

$$W_s(k) = D_s^{-1}(k) - A_s^{-1}(k)$$

is the wait in queue or idle time of the kth server to enter service.

From the above curves one can also evaluate the total waiting time of all customers or the total idle time of all servers, which are, respectively, the areas between $A_c(t)$ and $D_c(t)$, and between $A_s(t)$ and $D_s(t)$.

Of the four curves, described above, only $A_c(t)$ was considered as given. The others must be constructed from given values of the service times of the servers and given rules of service strategy, subject, however, to the constraint that $A_c(t) \geq D_c(t)$ and $A_s(t) \geq D_s(t)$.

The relation between the two curves $A_s(t)$ and $D_s(t)$ is determined entirely by the service times. If one has already constructed the curve $D_s^{-1}(k)$ for $k \leq k_0$; i.e., up to the time when the k_0^{th} service starts, then from given values of the service times S_k of the kth service, one can evaluate a set of future service completion times

$$D_s^{-1}(k) + S_k \quad , \quad k \leq k_0 , \qquad (2.3)$$

each of which will also represent a time at which a server becomes available, thus a step in the curve $A_s(t)$.

If the S_k are all equal, $S_k = s$ for all k , then these numbers will be monotone non-decreasing in k as is $D_s^{-1}(k)$. The servers will be used in cyclic order so that the kth server to complete service will become the $(k + n)^{th}$ server to become available. The time the $(k + n)^{th}$ server becomes available is the time that the kth service was completed; i.e.,

$$A_s^{-1}(k + n) = D_s^{-1}(k) + s . \qquad (2.4)$$

Geometrically, the interpretation of (2.4) is that A_s is obtained from D_s

by a translation of the curve D_s horizontally by s and vertically be n as shown in Fig. I.3a. Equivalently,

$$A_s(t + s) = D_s(t) + n .$$ (2.4a)

A section of the curve, $D_s^{-1}(k)$ for $k \leq k_0$ or $D_s(t)$ for $t \leq D_s^{-1}(k_0)$ determines the curve $A_s^{-1}(k)$ up to $k = k_0 + n$ or $A_s(t)$ up to $t = D_s^{-1}(k_0) + s$. For $t < s$, $A_s(t) = n$ if all n servers were available initially.

If the S_k are not equal, but the $D_k^{-1}(k) + S_k$ are still monotone in k , then it is still true that the kth server to complete service becomes the $(k + n)^{th}$ available server and therefore

$$A_s^{-1}(k + n) = D_s^{-1}(k) + S_k .$$ (2.5)

One need only plot the graph of $D_s^{-1}(k) + S_k$ and translate it up by n to obtain $A_s(t)$. But if $D_s^{-1}(k) + S_k$ is not monotone in k , its curve, shown in Fig. I.3b, may be quite irregular and not produce a single-valued function of t . If the service times are bus trips, this means that buses have passed each other; i.e., the service behavior is not FIFO.

We are assuming here that all servers are equivalent, at least in the sense that the strategy of use depends only upon the number of servers available, not on their identity or past history. We are not interested in the curve $D_s^{-1}(k) + S_k$ itself, but only in the ordered set of times at which servers become free. These can be obtained graphically by projecting the values of $D_s^{-1}(k) + S_k$ onto the t-axis and renumbering them. The curve $A_s(t)$ is now drawn as a step function starting at height n with a unit step at each of these times, as shown in Fig. I.3b.

In the case where (2.5) applied, the values of $D_s^{-1}(k)$ for $k \leq k_0$ determine $A_s^{-1}(k)$ for $k \leq k_0 + n$, n steps higher. Any iterative scheme for construction of these curves can generally proceed in steps of n servers at a time. Otherwise, however, when (2.5) is not true, one may be able to move only one step at a time. If one knows $D_s^{-1}(k)$ for $k \leq k_0$ one can certainly evaluate $A_s^{-1}(k_0 + 1)$ because it must be less than $D_s^{-1}(k_0 + 1)$; the $(k_0 + 1)^{th}$ service cannot start until it is available. But $A_s^{-1}(k_0 + 1)$ must also be either 0 , or one of the $D_s^{-1}(k) + S_k$

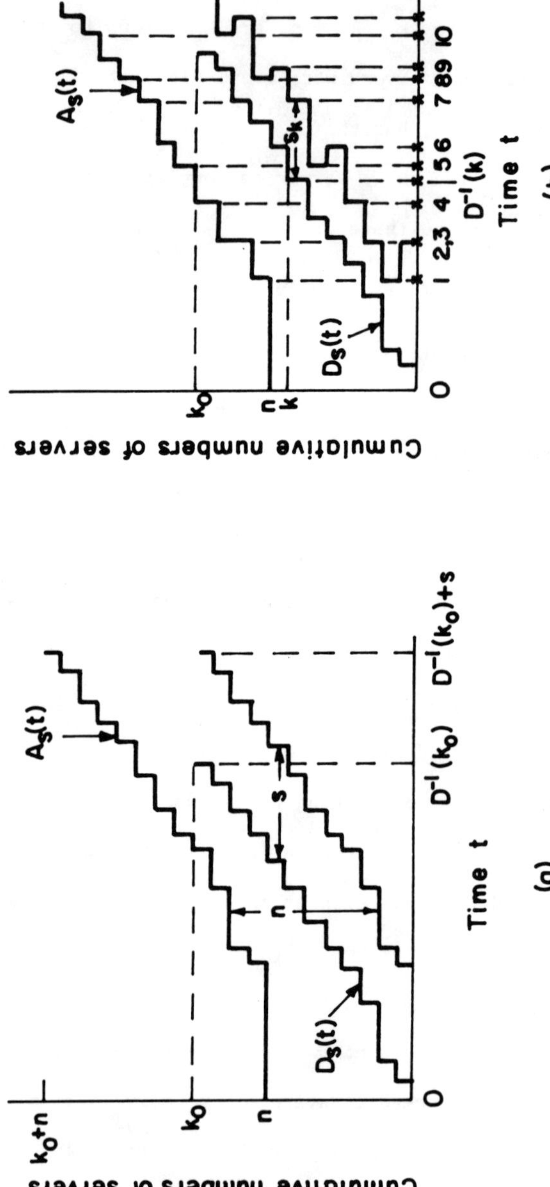

Fig. I.3 - Construction of cumulative arrival curve of servers $A_S(t)$ from cumulative curve of servers used, $D_S(t)$; (a) for $S_k = s$, (b) general S_k .

for some k , in particular some $D_s^{-1}(k) + S_k$ with $k \leq k_0$ since any others would be greater than $D_s^{-1}(k_0 + 1)$.

From $D_s^{-1}(k)$ for $k \leq k_0$, one will also know the curve $A_s(t)$ at least until time $D_s^{-1}(k_0)$, since no subsequent value of $D_s^{-1}(k) + S_k$ for $k > k_0$ can produce steps in $A_s(t)$ for $t < D_s^{-1}(k_0)$. This will usually determine $A_s^{-1}(k)$ up to some value of $k > k_0 + 1$. There is no guarantee, however, that the value of $A_s^{-1}(k_0 + 2)$ will be determined. If the $(k_0 + 1)^{th}$ server to become available is also the next one to enter service (all servers are busy), and its service time is so short that the $(k_0 + 1)^{th}$ server passes all the others in the service, then $A^{-1}(k_0 + 2)$ would be $D_s^{-1}(k_0 + 1) + S_{k_0+1}$.

The curves $D_c(t)$ and $D_s(t)$ are constructed from the curve $A_c(t)$ and the service strategy. If, at any time, either the queue of customers or available servers is zero, no customer will enter service, i.e., $D_c(t)$ remains unchanged, neither will a server enter service, i.e., $D_s(t)$ remains unchanged (except in the case where the strategy permits sending an empty vehicle on a trip, a service to zero customers). If, however, both a server and a customer are available (or have just become available), the strategy will specify whether or not a server will be used at that moment, i.e., whether or not $D_s(t)$ has a unit step. If so, the strategy will also specify how many customers enter service at the same time; i.e., the height of the step in $D_c(t)$.

Generally, this whole procedure of constructing the curves $D_s(t)$, $D_c(t)$, and $A_s(t)$ from $A_c(t)$ must be done sequentially because at any time t , the behavior of $D_s(t)$ and $D_c(t)$ will depend on the value of $A_s(t)$, which depends upon the previous values of $D_s(t)$.

Although we have, so far, been building a framework suitable for the analysis of bulk service servers, most of the following analysis will be devoted to the special case in which each server serves only one customer at a time, or a nearly equivalent bulk-service system for which the strategy requires that the server must serve in batches of the same number each time.

If a server serves only one customer at a time, the relation between the four curves $A_c(t)$, $A_s(t)$, $D_c(t)$, and $D_s(t)$ is relatively simple. Since the number of customers to enter the service must always be the same as the number of servers to enter service,

$$D_c(t) = D_s(t) \equiv D(t) \qquad (2.6)$$

are the same curves. Furthermore, one will never allow both a queue of customers and of servers to exist simultaneously. Therefore either

$$D_c(t) = A_c(t) \ , \ N_c(t) = 0$$

or

$$D_s(k) = A_s(t) \ , \ N_s(t) = 0 \ .$$

But since $D_c(t) \leq A_c(t)$ and $D_s(t) \leq A_s(t)$ for all t , it follows that

$$D(t) = \min \ [A_c(t) \ , \ A_s(t)] \ . \qquad (2.7)$$

Equivalently, the time at which k servers have entered service is equal to the time at which k customers have entered service

$$D_c^{-1}(k) = D_s^{-1}(k) \equiv D^{-1}(k) \ , \qquad (2.6a)$$

but this time must be either the time when the kth server was available or the kth customer arrived, i.e., either $D_s^{-1}(k) = A_s^{-1}(k)$ or $D_c^{-1}(k) = A_c^{-1}(k)$. Therefore

$$D^{-1}(k) = \max \ [A_c^{-1}(k) \ , \ A_s^{-1}(k)] \ . \qquad (2.7a)$$

This relation between D and A_s , along with the previously described procedure for constructing $A_s^{-1}(k)$ from the $D_s^{-1}(k) + S_k$, will determine both A_s and D . This must generally be done iteratively. If we know $D^{-1}(k)$ for $k \leq k_0$, then we can always determine at least $A_s^{-1}(k_0 + 1)$, and, in the case $S_k = s$, the $A_s^{-1}(k)$ up to $k = k_0 + n$. Equation (2.7a) will then determine (at least) the next value of $D^{-1}(k)$, $D^{-1}(k_0 + 1)$.

For $S_k = s$, this construction is very simple and can be done n servers (or customers) at a time. This is illustrated in Fig. I.4, with the integer steps in the curves smoothed out. Starting at time $t = 0$, we know that $A_s(t) \geq n$ and therefore $D_s(t) = A_c(t)$ is known at least until $D(t) = n$. This allows us to start the construction of $A_s(t)$, which, for $S_k = s$, is obtained by simply

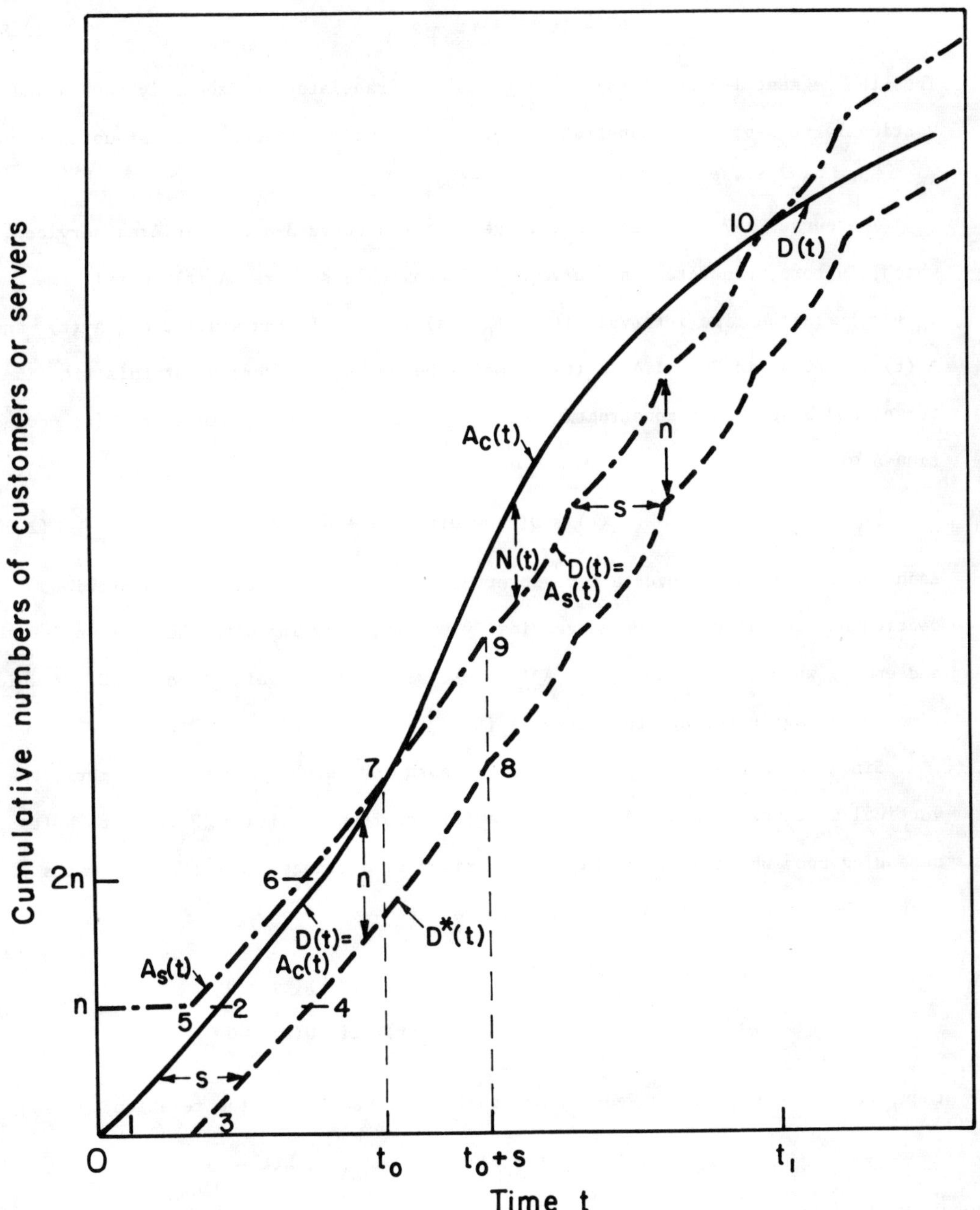

Fig. I.4 - Construction of $A_S(t)$ and $D(t)$ from $A_C(t)$ for $S_k = s$.

translating $A_c(t) = D(t)$ horizontally by s and vertically by n

$$A_s(t) = A_c(t - s) + n \; . \tag{2.8}$$

Thus the segment 1-2 of $A_c(t)$ in Fig. I.4 is translated horizontally to 3-4 and vertically to 5-6. This construction of $A_s(t)$ will continue at least until time t_0 where $A_s(t) = A_c(t)$ for the first time.

The curve 1-7 of $D(t) = A_c(t)$ generates a curve 3-8 of completed services, which, in turn, generates the curve 5-9 of available servers $A_s(t)$ until time $t_0 + s$. In the time interval $(t_0 , t_0 + s)$, $D(t)$ is the smaller of $A_s(t)$ and $A_c(t)$. If, as in Fig. I.4, $A_s(t)$ remains below $A_c(t)$ throughout this time interval and beyond, the construction of $A_s(t)$ from previous values of $D(t)$ continues with

$$A_s(t) = D(t) = D(t - s) + n \; . \tag{2.9}$$

Each section of $D(t)$ over a time interval s is a displacement of a previous section horizontally by s and vertically by n , starting with the section 7-9 and ending when the curve $D(t) = A_s(t)$ crosses $A_c(t)$ again at point 10, at which time $D(t)$ again follows the curve $A_s(t)$.

Since the curves of Fig. I.4 describe both the servers and the customers, the vertical distance between $A_s(t)$ and $A_c(t)$ represents either $N_s(t)$ or $N_c(t)$ depending upon which curve is higher. Horizontal distances represent the waits of either the servers or the customers. If we let

$$N(t) = A_c(t) - A_s(t) = \begin{cases} N_c(t) & \text{if} \quad N(t) > 0 \\ -N_s(t) & \text{if} \quad N(t) < 0 \end{cases} \tag{2.10}$$

then, for $S_k = s$, one can easily see that $N(t)$ satisfies the relation

$$N(t) = A_c(t) - A_c(t - s) - n + \max [0 , N(t - s)] \; . \tag{2.11}$$

because, during the time $t - s$ to t , n servers will have been available, whereas the number of customers to be served is the number of new arrivals $A_c(t) - A_c(t - s)$ plus any residual queue that existed at time $t - s$. This equation relates $N(t)$ directly to $A_c(t)$ without all the other curves. We will use this later for some of the analytical estimations but continue to use the graphical representations to show the complete evolution of the system.

If the S_k are not equal, the graphical construction is more complicated because one must order the values of $D^{-1}(k) + S_k$ before constructing $A_s(t)$, but again there are two basic types of construction, one in which the $A_s(t)$ is evaluated from the past values of $D(t) = A_c(t)$ and the other in which it is evaluated from past values of $D(t) = A_s(t)$.

To analyse a system in which each server can serve batches of b customers at a time but the service strategy will not allow a server to start service until it is full, one should first construct from the given curve $A_c(t)$, a new curve $A_c'(t)$ representing the cumulative integer multiples of b customers that have arrived by time t . The curve $A_c'(t)$ is simply a step function with steps of height b as shown in Fig. I.5. If we interpret the curve $A_c'(t)$ as the arrival curve for the batch service, the system will behave exactly as if the service served only one "customer" at a time, but the customer is the batch of size b . Having evaluated queues, delays, etc., for the batches, one need only add to these the extra queues or delays corresponding to the distances between $A_c(t)$ and

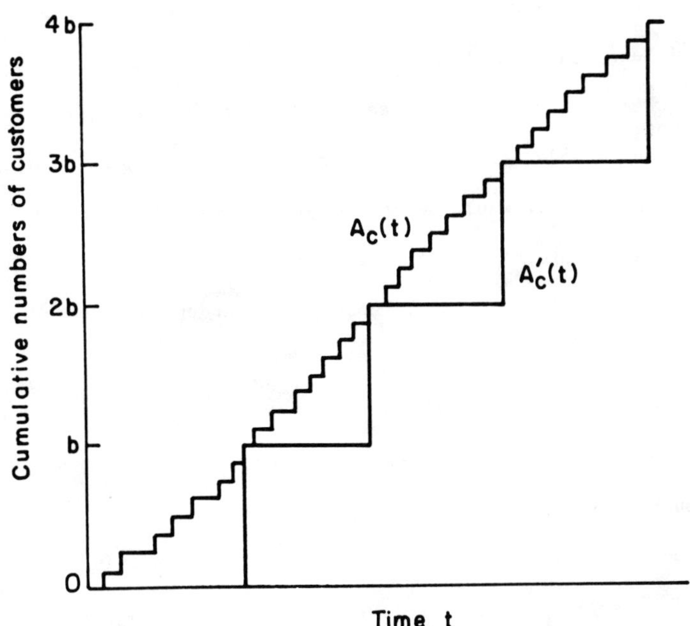

Fig. I.5 – Cumulative number of customers, $A_c(t)$ and cumulative number of batches of size b .

$A_c'(t)$. In the following, we will interpret the $A_c(t)$ to be the curve $A_c'(t)$ if we wish to consider batch service. It is to be understood that the additional delays of Fig. I.5 are to be added.

It would be desirable to systematically analyse the consequences of each of the bulk service strategies listed in the last section. Each has its own virtues in special situations. But our purpose here is to move on to some of the stochastic properties and methods of approximation. Hurdle[7] has already made a fairly systematic analysis of optimal dispatching strategies for a public transportation route with $n \gg 1$, using continuum approximations and a constant trip time s , but neglecting stochastic effects completely. He also made extensive use of graphical constructions very similar to those described above, with separate graphs for customers and servers. The above service strategy, which corresponds to dispatching vehicles only when full, was shown to be the optimal strategy while there was queueing of customers and for at least one service time s before queueing started. A more detailed analysis of this strategy, with stochastic effects, therefore, is of some practical interest. At other times, however, the optimal strategies of dispatching usually involved some compromise between providing good service immediately or saving vehicles to provide better service later.

There is one other strategy which would be relatively simple: use an available server whenever there is any queue of customers, even for a server of capacity $b > 1$. If b is moderately large (larger than 10, say) and the customer demand is close to capacity, it would be unlikely, however, that a server would have to wait because there were no customers to serve. The n servers would be serving all the time with a pattern that is nearly independent of the customers (in Fig. I.3, $A_s(t) = D_s(t)$ for nearly all t) . The problem would reduce to an analysis of the queueing of customers for a bulk server which provides service according to a specified pattern of time intervals between services. In most cases, the behavior of such a system would be nearly equivalent to a single server system with bulk service and "service time" (actually the time intervals between service) of $E\{S\}/n$.

3. Stochastic Properties

If a server serves only one customer (or batch of size b), then, for any given values of the S_k and $A_c(t)$, particularly if $S_k = s$, the graphical construction of $D(t)$ and $A_c(t)$ can be done very quickly. To treat the problem with stochastic S_k and $A_c(t)$, we could still consider the same graphical construction as in section 2, but interpret the $A_c(t)$ and S_k as a single realization of the system. We could (at least conceptually) imagine all possible curves $A_c(t)$ and values of the S_k along with the corresponding curves $D(t)$ and $A_s(t)$, queue lengths, waits, etc., and then average the latter over a suitable sample of curves $A_c(t)$ and sets of S_k consistent with some given probability distributions of $A_c(t)$, S_k , to obtain estimates of expectations.

To formulate this analytically and evaluate expected queue lengths exactly from probability distributions of $A_c(t)$ and S_k appears to be hopelessly complicated. Since this is the procedure most theorists would likely wish to follow, if they knew how, it is not surprising that the problem has been avoided. One could analyse the problem by simulation, however (either by hand or computer). Even about 10 "typical" realizations would give some qualitative estimates accurate enough for most practical applications (experimentally observed variances frequently deviate from any reasonable theoretical models by as much as a factor of two anyway). A computer analysis of a few thousand realizations could yield quite accurate estimates of the probability distributions of queue lengths at every value of t .

Whereas purely analytic procedures given formulas that cannot be evaluated, simulations give numerical values for only one problem at a time and are rather clumsy for exploring the qualitative consequences of various possible system designs (which is typically the ultimate goal). Here we would like to follow an intermediate procedure in which we imagine the possible consequences of the simulations and try to make crude analytic estimates only of those things which seem to be most relevant.

There is a wide variety of possible shapes for $A_c(t)$ and a wide range of possible distributions of the S_k . We will not try to obtain formulas that apply to all cases simultaneously. We will first try to identify problems which can be solved easily (at least approximately) and gradually work our way into the more difficult cases.

Since we assume that $n \gg 1$, and consequently we are dealing with large counts of customers or servers, there will be extreme situations in which stochastic effects are negligible in the sense that the standard deviation of $N(t)$, say, is small compared with $E\{N(t)\}$ or n . In such cases we may be able to evaluate $N(t) \simeq E\{N(t)\}$ by simply replacing the various realizations of $A_c(t)$ by a single curve for $E\{A_c(t)\}$, and, if S_k is also a random variable, replacing it by some suitable constant, perhaps $E\{S\}$. There may also be situations in which the fluctuations are not negligible in the above sense, but, nevertheless, because of the linearity of certain mathematical operations, the value of $E\{N(t)\}$ can be found by the same method of replacing stochastic variables by appropriate averages.

Although we will postpone making quantitative estimates of variances of $N(t)$ and related things until we see where this is necessary, we will at all times have in mind some notion of their order of magnitude based upon certain postulated properties of $A_c(t)$ and the S_k .

The mean time interval between customer arrivals will always be considered small compared with any other relevant time constants, therefore, during any significant time interval t to $t + \tau$, the expected number of arrivals $E\{A_c(t + \tau) - A_c(t)\}$ will be large compared with 1. We also assume that if statistical dependencies exist between the arrival times of successive arrivals, they decay within a few interarrival times (as with a Poisson process, a renewal process with finite moments, a clustered Poisson process, etc.) so that the number of arrivals in non-overlapping time intervals (of sufficient duration) are nearly independent. The variance of counts add in the same way as the expectations and

$$\frac{\text{Var}\{A_c(t + \tau) - A_c(t)\}}{E\{A_c(t + t) - A_c(t)\}} \equiv I(t , \tau) \simeq I(t) \qquad (3.1)$$

should be nearly independent of the time τ if τ is large compared with the mean interarrival time. We will also assume that $I(t) = I$ is independent of t . The latter assumption does not seem very critical to any of the analysis to follow, but to include a time dependence of I does not appear to be a very significant generalization.

If $A_c(t)$ is generated by a Poisson process (stationary or not), then $I(t, \tau)$ is exactly 1 for all t and τ . For the processes that we have in mind (cars on a highway, people arriving at a bank or bus depot, or people arriving late for appointments, etc.) we would expect I to be less than 2 or 3 . For a service in which each server serves single customers, however, we would not normally expect I to be much less than 1 except when there is some underlying schedule of arrivals (appointments). Note that this hypothesis will usually exclude the possibility that the arrivals were generated as the output of some other service facility of the type being considered, which will cause statistical dependencies on a time scale comparable with the mean service time $E\{S\}$.

In dealing with batch service, the "customer" is interpreted as a batch of b customers. If the above hypothesis is true of arrivals of individual customers, then during any time τ covering several batches, the expected number of batches is 1/b times the expected number of customers, and the variance of the number of batches is $1/b^2$ times the variance of the number of customers. Thus the value of I for batches is 1/b times that of individual customers. In treating batches of large size, the effective value of I may be small compared with 1 , comparable with 1/b . For $b \gg 1$, the arrivals of batches behave nearly like regular arrivals.

The number of arrivals during a mean service time will be important in much of the following analysis. When the system is operating close to capacity this number will be comparable with the number of servers. The variance will be about In , the standard deviation about $(In)^{1/2}$, and the fractional fluctuations of order $(I/n)^{1/2}$. The methods to be used are based upon treating $(I/n)^{1/2}$ as a small quantity, although actually, for the situations we have in mind, n is in the range of 10 to 20 , $n^{1/2}$ in the range of 3 to 5 , so that $n^{-1/2}$ itself is not very small. One cannot usually neglect fluctuations completely, but one can esti-

mate their properties by asymptotic approximations. If the server is an elevator
or bus that serves b = 10 to 50 customers at a time, however, $(I/n)^{1/2}$ will be
small (less than 1/10).

When, later on, we need to know an approximate probability distribution of the
number of arrivals, we will further assume that these counts are approximately normally distributed with the appropriate mean and I value. This should be a reasonable assumption in most applications if the mean count is large compared with 1
(or more specifically if the standard deviation is moderately large, at least 2 or
3).

For the service times we will assume that the S_k are (nearly) independent
identically distributed random variables with a distribution function

$$G(\tau) = P\{S_k \le \tau\} , \quad \overline{G}(\tau) = 1 - G(\tau) = P\{S_k > \tau\} , \qquad (3.2)$$

also that the coefficient of variation of S

$$C_S = [\text{Var}\{S\}/E^2\{S\}]^{1/2} \qquad (3.3)$$

is finite with a value less than about 1 or 2. For some examples, such as trip
times of buses we can expect a small value of C_S (maybe 1/10), in other cases we
might expect S to be exponentially distributed with $C_S = 1$.

It might seem that under these hypotheses, the numbers $D_s^{-1}(k) + S_k$ of (2.3)
and their corresponding representations in Fig. I.3b would have a variance due to
the S_k which would completely dominate any fluctuations caused by the $A_c(t)$.
However, the variances of these numbers themselves are not of importance, it is the
variance of $A_s(t)$ or the ordered values of the $D_s^{-1}(k) + S_k$ that are important.
A large variance of S_k will not necessarily cause a large variance of $A_s(t)$.

We will see later that there is a rather complicated interplay between the
fluctuations in the $A_c(t)$ and the S_k if, as a result of these fluctuations, it
is uncertain as to whether or not a queue of customers or servers exists. The point
process of service completions is, however, the superposition of the processes of
service completions by each of the n servers. If each server is kept busy most
(or all) of the time, the process of service completions by each server is approximately (exactly) a renewal process. Although the renewal processes are not statis-

tically independent, but correlated through their dependence upon some past arrival times of customers, the ratio of the variance to mean of the number of counts will still be comparable with 1 under the above hypotheses. Suffice it to say, for now, that fluctuations in queue lengths caused by the variance of the S_k can be considered to be of the same order of magnitude as those due to the $A_c(t)$. Their contributions may be additive or they might even cooperatively cancel because of their different roles.

For random S_k , the graphical representations are rather tedious to construct and even more difficult to interpret. We shall use the graphs mostly to show what happens in the special case $S_k = s$ but random $A_c(t)$, and then use a mixture of graphical and analytic techniques to extend this to the case of random S_k .

The above assumptions guarantee that the relative fluctuations in arrivals or departures, $A_c(t)$ and $D(t)$, are small compared with their expectations, but do not, in themselves, guarantee that the fluctuations in the queue, $N(t)$, are small compared with the expected number of customers in the system $E\{N(t)\} + n$, or n , or $|E\{N(t)\}|$ or $|E\{N(t)\}| + n$. But since $N(t)$ is the difference between (dependent) arrivals and departures, a relatively large variance for $N(t)$ can arise only if the mean number of arrivals and departures nearly cancel, but their variances do not. To obtain a standard deviation for $N(t)$ of the order of magnitude n , for example, one must have had at least of the order of n^2 arrivals and departures. In terms of a diagram such as Fig. I.4, this means that on a graph with a scale of counts at least of order n , various realizations $A_c(t)$ will be nearly coincident (for sufficiently large n) . In this case, fluctuations would be significant only if one were concerned with quantities (particularly $N(t)$) , the values of which could not be easily measured on a graph drawn to such a scale. To see or measure the relevant quantities, one would need to magnify certain portions of the graph. As a practical matter, however, for n about 10 or 20, this is not really a problem; one can see and measure fluctuations on a graph with many realizations. But it is helpful conceptually, in identifying the order of magnitudes of various quantities relative to various powers of n, to imagine what would happen for arbitrarily large n .

CHAPTER II - APPROXIMATION METHODS

1. Introduction

In Chapter I*, we described some methods of representing graphically the evolution of realizations of queueing systems containing multiple servers. We described some of the postulates to be made about arrival processes and service times, and gave some hints as to how we would approach the analysis of a system of n servers with $n \gg 1$.

Here we will describe various types of approximations that can be employed to analyse the behavior of an n-server system for which each server serves only one customer at a time (or batches of a specific number b of customers). In general, it will be assumed that the arrival rate of customers is time-dependent with an expected cumulative arrival curve $E\{A_c(t)\}$ typical of a rush hour; the arrival rate

$$\lambda(t) = dE\{A_c(t)\}/dt \qquad (1.1)$$

increases monotonically to a maximum and then decreases in a manner similar to that shown in Fig. I.4 for a single realization of $A_c(t)$.

If the arrival rate becomes sufficiently large as to cause queueing of customers (for all or some realizations), it is obvious from Fig. I.4 that one should decompose the analysis into several parts, treating separately the behavior of the system (1) prior to the time that queueing occurs, (2) during possible "transition periods" when queueing may occur for some, but not all, realization, and (3) during the time (if any) when a queue exists for nearly all realizations of the $A_c(t)$ and S_k . Particularly, the latter two must be further decomposed into various types of behavior depending mostly upon how the duration of these states compares with the expected service time $E\{S\}$.

In section 2, we analyse the behavior of the system during times when it is virtually certain that there is no queue of customers (in section I-2, $N(t) < 0$, $N_c(t) = 0$, $N_s(t) > 0$) . This is mostly a review of known results since, under these conditions, the system behaves approximately as would a system of infinitely many servers. In sections 3-5, we treat the case in which queueing does occur

---

*Equations, sections, and figures of Chapter I will be identified by I(·).

but the service time is very large, comparable with the duration of the rush hour. The method here is mostly graphical and exploits the fact that the curve $A_s(t)$ of cumulative available servers is determined by how servers were used at times of the order of a mean service time earlier, perhaps before queueing started. Section 3 deals with constant service time $S_k = s$; section 4 with random service time but small variance; and section 5 with large variance.

Cases of "small" service times will be treated in Chapter III.

2. Approximations-- no customer queue

The analysis of the system during times when there is no queue of customers is analytically quite straightforward and well known. As a prelude, however, to some of the methods to be used later, it is convenient to analyse the behavior in the context of the graphical constructions of realizations discussed in Chapter I, and otherwise use somewhat unconventional interpretations. We will first look at the nearly trivial case of fixed service time $S_k = s$.

In (I 2.10) we defined $N(t)$ as the number of customers in queue, $N_c(t)$, if $N(t) > 0$, or the negative of the number of idle servers, $-N_s(t)$, if $N(t) < 0$. The number of customers in the system (queue plus service) is $N(t) + n$. If $N(t) < 0$, $N(t) + n$ is also equal to the number of busy servers and to the number of customers in service. In the present case, this is a rather clumsy notation, because $N(t) + n$ is the quantity of primary interest (rather than $N(t)$ itself), and it does not depend upon n as long as $N(t)$ stays negative, i.e., $N(t) + n < n$. We will suffer with this temporarily, however, rather than introduce new notation; we will eventually be using $N(t)$ more often than $N(t) + n$.

For $S_k = s$, it follows from (I 2.11) that

$$N(t) + n = A_c(t) - A_c(t - s) \qquad (2.1)$$
$$= \text{number of arrivals during } (t - s \text{ , } t)$$

provided that there is no queue of customers at time $t - s$, $N(t - s) < 0$.

In relation to some of the arguments in Section I.3, it is important to observe first that $N(t) + n$ depends upon the arrivals only during a finite period of time s . Even though we might have drawn $A_c(t)$ in Fig. I.4 over a long time span (large compared with s), the fluctuations in $A_c(t)$ might have been large compared with

$N(t)$, and the magnitude of $N(t)$ might be difficult to measure on the graph, we could have drawn the graph on a finer scale starting at time $t - s$ as a new time origin for counting arrivals (regardless of the random count $A_c(t - s)$ relative to some fixed origin) and with a time scale of order s . On the latter graph we would see that the relative fluctuations in $N(t) + n$ are small. According to the postulates of Section I.3.

$$\sigma_N \equiv [\text{Var}\{N(t) + n\}]^{1/2}$$
$$= [\text{Var}\{N(t)\}]^{1/2} \simeq I^{1/2}[E\{N(t) + n\}]^{1/2} , \qquad (2.2)$$

and $N(t) + n$ should have approximately a normal distribution. If $E\{N(t)\} + n$ is of order n , the relative fluctuation in $E\{N(t) + n\}$ is of order $\sigma_N/n \simeq (I/n)^{1/2}$, which we interpret as being small compared with 1.

To evaluate $E\{N(t)\} + n$, it is not necessary, however, to subtract away the fluctuations in $A_c(t - s)$ by choosing this as a new origin for counting. Because of the linearity of (2.1)

$$E\{N(t)\} + n = E\{A_c(t) - A_c(t - s)\} = E\{A_c(t)\} - E\{A_c(t - s)\} , \qquad (2.3)$$

which implies that (with $N(t - s) < 0$) , the value of $E\{N(t)\}$ could be found directly from a graph of $E\{A_c(t)\}$. Furthermore, if there is no customer queue at time t (for $S_k = s$ or not), from (I 2.7)

$$D(t) = A_c(t) \quad \text{implies} \quad E\{D(t)\} = E\{A_c(t)\} \qquad (2.4)$$

and for $S_k = s$ (with or without queueing), from (I 2.9)

$$A_s(t) = D(t - s) + n \quad \text{implies} \quad E\{A_s(t)\} = E\{D(t - s)\} + n . \qquad (2.5)$$

Thus $E\{N(t)\}$ can be evaluated from the curve of $E\{A_c(t)\}$ by the same graphical construction as used to find $N(t)$ in Fig. I.4 from any realization $A_c(t)$, regardless of the magnitude of the fluctuations, as long as they do not cause a queue of customers.

The above is not completely trivial. One cannot determine the expected idle time of a server from the curve $E\{A_c(t)\}$ because the notation implies that this is an expectation of $A_c(t)$ for fixed t . If one wants the expected idle time of the kth server to enter service $W_s(k)$, one must take expectations for fixed k .

Under the same conditions as above

$$W_s(k) = D^{-1}(k) - A_s^{-1}(k) = A_c^{-1}(k) - A_c^{-1}(k - n) - s \ .$$

By analogous arguments, $E\{W_s(k)\}$ could be obtained from a graph of $E\{A_c^{-1}(k)\}$. The curve for $E\{A_c^{-1}(k)\}$ is not the same as the inverse of the curve for $E\{A_c(t)\}$, although in most practical applications, the difference between these should not be enough to make much difference.

If $A_c(t)$ is an experimentally observed realization, its curve is likely to have some wiggles which would not be reproduced by other realizations (they are stochastic). In many practical situations one would find, or could assume, that $E\{A_c(t)\}$ is a smooth function on some time scale large compared with the time between arrivals. If this is the case, one can often estimate $E\{A_c(t)\}$ from a single realization $A_c(t)$ by simply drawing a smooth curve approximation to $A_c(t)$ (or $A_c^{-1}(k)$) , averaging out wiggles of magnitude $[I \, A_c(t)]^{1/2}$.

One could also evaluate $N(t)$ from a single realization. If $N(t) + n$ behaves as if it had fluctuations of order $\sigma_N = I^{1/2}[N(t) + n]^{1/2}$ superimposed on a smooth curve, one could again just smooth out some of the fluctuations to obtain an estimate of $E\{N(t)\} + n$. Since $N(t) + n$ is a count of arrivals over a time interval s , it is itself a "running average" of differences in $A_c(t)$. If there were some distortion in the curve $A_c(t)$, its effect on $N(t)$ would persist for a time of order s , i.e., until the customers that caused the distortion have left the service. Stochastic fluctuations in $N(t)$ should, therefore, not only be of order σ_N but should rise and decay on a time scale of order s (the latter will have some important consequences when queueing starts).

The above behavior of $A_c(t)$ or $N(t)$ could, in principle, be checked by obtaining several experimental realizations, comparing them, and averaging them, but repetition of experiments is often tedious or impossible to do under "identical conditions." Whether the above procedure is proper or not, this is what is usually done: the "average" behavior is inferred from a smoothing of a single realization and the "stochastic" properties are inferred from the magnitude of the deviations of the realization from the smooth curve rather than from "repetition of the experiment under identical conditions." We mention this here because in the following

analysis we will try, at all times, to bias the methods of analysis so as to minimize the amount of data needed as input rather than to maximize the accuracy, given
perfect data. Typically one can obtain reasonable estimates of $E\{A_c(t)\}$ from observations but only very crude estimates of stochastic behavior.

If $E\{A_c(t)\}$ is a smooth function, we can define an arrival rate (1.1), (perhaps as the slope of a smoothed realization of $A_c(t)$), from which (2.3) becomes

$$E\{N(t)\} + n = \int_{t-s}^{t} \lambda(\tau)d\tau .$$
(2.6)

If $\lambda(\tau) = \lambda$ is constant, then

$$E\{N(t)\} + n = s\lambda .$$
(2.6a)

Since $E\{N(t)\} + n$ is the expected number of busy servers or customers, this is a
special case of the Little formula [8] "$L = \lambda W$."

If $\lambda(\tau)$ is slowly varying so as to be approximated by a linear function over
the time range $(t - s , t)$, then (2.6) can be approximated by

$$E\{N(t)\} + n \simeq s\lambda(t - s/2)$$
(2.6b)

which reflects the fact that the number of customers in service now is determined by
the arrival rate at an earlier time (approximately $s/2$ earlier).

The above formulas were based upon the hypothesis that there was no queue of
customers during the time period in question. Since fluctuations in $N(t)$ are expected to be of order σ_N , and the distribution of $N(t)$ to be approximately
normal with this as its standard deviation, we expect these formulas to be approximately correct if $E\{N(t)\}$ plus any fluctuation (say two standard deviations) is
almost always negative, i.e.,

$$- E\{N(t)\} \gtrsim 2 \sigma_N \simeq 2(In)^{1/2} .$$
(2.7)

If $\lambda(t)$ is slowly varying, (2.7) implies

$$1 - (s/n)\lambda(t - s/2) \gtrsim 2(I/n)^{1/2} .$$
(2.8)

The quantity $s\lambda(t)/n$ is customarily called the traffic intensity,

$$\rho(t) = s\lambda(t)/n .$$
(2.9)

If $\rho(t)$ stays below 1, the queues (if any) stay finite, but if $\rho(t)$ stays above 1, the queue of customers will grow with time.

One of the most important consequences of the assumption $n \gg 1$ is that queueing does not become a significant issue until the traffic intensity becomes "quite close" to 1, within about $2(I/n)^{1/2}$ of 1. For a toll booth with $I = 1$, $n \sim 20$, $\rho(t)$ close to 1 means greater than about 0.5 or 0.6, but for 10 buses of 40 people each, $n = 10$, $I \sim 1/40$, it means greater than about 0.9

If, for a rush hour, $\lambda(t)$ is an increasing function of time prior to its peak, then (2.6b) will hold at least until (2.8) fails. One should observe, however, that (2.6) or (2.6b) is correct at time t provided only that there was no customer queue at time $t - s$, even though a queue may form during the time $(t - s , t)$.

Although the above formulas were based upon the assumption that $S_k = s$ for all customers, most of these results have simple generalizations to the case of random service times.

Since the relation between $D_s(t)$ and $A_s(t)$ described in section I.2 does not depend upon the customer arrivals, we will, for future reference, imagine that servers (or customers) enter service at a rate $\mu(t)$.

$$\mu(t) \equiv dE\{D_s(t)\}/dt . \tag{2.10}$$

In the present situation, with no queueing and one customer per server $D(t) = D_s(t) = A_c(t)$, and so $\mu(t) = \lambda(t)$ is also the arrival rate of customers.

To evaluate $E\{A_s(t)\}$, which is n plus the expected number of service completions, we note that the expected number of servers (customers) to enter service during the time interval $t - \tau$ to $t - \tau + d\tau$ is $\mu(t - \tau)d\tau$. If a kth server enters service at time $t - \tau$, it will have left service by time t if $S_k \leq \tau$. The expectation of leaving is $G(\tau)$. Thus

$$E\{A_s(t)\} = n + \int_0^\infty d\tau\mu(t - \tau)G(\tau) = n + \int_0^\infty dG(\tau)E\{D(t - \tau)\} . \tag{2.11}$$

i.e., $E\{A_s(t)\} - n$ is a weighted average of horizontal translations of $E\{D(t)\}$.

Correspondingly, the expected queue of idle servers is

$$E\{N_s(t)\} \;=\; E\{A_s(t)\} - E\{D(t)\}$$

$$= \; n - \int_0^\infty d\tau \mu(t-\tau)\overline{G}(\tau) \tag{2.12}$$

$$= \; n - \int_0^\infty dG(\tau)E\{D(t) - D(t-\tau)\} \; .$$

With no queueing (prior to time t) , $\mu(t-\tau) = \lambda(t-\tau)$ and $D(t-\tau) = A_c(t-\tau)$, this gives

$$E\{N(t)\} + n \;=\; \int_0^\infty d\tau \lambda(t-\tau)G(\tau)$$

$$= \int_0^\infty dG(\tau)E\{A_c(t) - A_c(t-\tau)\} \; . \tag{2.13}$$

which for $S_k = s$

$$G(\tau) = \begin{cases} 0 & \text{for } \tau < s \\ 1 & \text{for } \tau \geq s \end{cases} \tag{2.14}$$

reduces to (2.3) or (2.6).

Most of the qualitative properties of $N(t)$ discussed above for $S_k = s$ are still true even though the formulas are changed. Despite the possibility that $A_c(t)$ may have large fluctuations from one realization to the next, $N(t)$ depends upon the arrivals only during a restricted range of earlier times, within a time of order $E\{S\}$ earlier. The fluctuations in $A_c(t)$ partially cancel in the evaluation of $N(t)$.

It is no longer possible to measure $E\{N(t)\}$ directly from a graph of $E\{A_c(t)\}$, but one can compute, from $E\{A_c(t)\}$ and $G(\tau)$, a graph of $E\{A_s(t)\}$ using (2.11) with $D(t-\tau) = A_c(t-\tau)$ or $\lambda(t-\tau) = \mu(t-\tau)$. This may require a numerical integration of (2.11) unless these functions have some simple analytic form. Since (2.11) shows that $E\{A_s(t)\}$ is a weighted average of past values of $E\{A_c(t-\tau)\}$, we would expect $E\{A_s(t)\}$ to be a smoother curve than $E\{A_c(t)\}$.

If $\lambda(t) = \lambda$ is constant, then (2.13) gives

$$E\{N(t)\} + n \;=\; \lambda \int_0^\infty d\overline{G}(\tau) = \lambda E\{S\} \tag{2.15}$$

as the generalization of (2.6a), again a well-known special case of the formula

"$L = \lambda W$." This shows that if one can approximate a random service time by an "equivalent" constant time, the proper choice would be $E\{S\}$. We use this to re-define the traffic intensity as

$$\rho(t) = E\{S\}\lambda(t)/n \ .\tag{2.16}$$

If $\lambda(t)$ is slowly varying so as to be approximated by a linear function over time intervals covering most of the range of S , i.e.

$$\lambda(t - \tau) \simeq \lambda(t) - \tau d\lambda(t)/dt \ ,\tag{2.17}$$

then (2.13) gives

$$E\{N(t)\} + n \simeq \lambda(t)\int_0^\infty d\tau\overline{G}(\tau) - [d\lambda(t)/dt]\int_0^\infty d\tau\tau\overline{G}(\tau)$$

$$\simeq \lambda(t)E\{S\} - [d\lambda(t)/dt]E\{S^2\}/2$$

or

$$E\{N(t)\} + n \simeq E\{S\}\lambda\left(t - \frac{E\{S^2\}}{2E\{S\}}\right) \ .\tag{2.18}$$

This is the generalization of (2.6b). Again we see that $E\{N(t)\}$ depends upon the arrival rate at an earlier time, but now the time displacement is

$$\frac{E\{S^2\}}{2E\{S\}} = \frac{E\{S\}}{2}\left[1 + c_{\underline{S}}^{\overline{2}}\right] \geq \frac{E\{S\}}{2} \ ,\tag{2.19}$$

where C_S is the coefficient of variation of S , (I 3.3). Thus for fixed $E\{S\}$, an increase in Var $\{S\}$ tends the delay the response of the system to slow changes in $\lambda(t)$, twice as much for an exponentially distributed S as for fixed S . Of course, (2.8) cannot be represented in terms of just a single equivalent constant service time.

The quantity (2.19) occurs quite frequently in the theory of stationary point processes[4] . If $G(\tau)$ is the distribution function of the time between events, (2.19) is the mean time to the first event starting from a randomly chosen time origin. Specifically, if we let S_0 be a random variable with a probability density $\overline{G}(\tau)/E\{S\}$, distribution function

$$G_0(\tau) \equiv \int_0^\tau dx \overline{G}(x)/E\{S\} \,, \tag{2.20}$$

then

$$E\{S_0\} = \frac{1}{2} E\{S^2\}/E\{S\} \tag{2.20a}$$

and (2.18) becomes

$$E\{N(t)\} + n \simeq E\{S\}\lambda(t - E\{S_0\}) \,. \tag{2.18a}$$

The estimates of the fluctuations in $N(t)$, which restricts the range of validity of these formulas, are somewhat more complicated than before. As long as there is no queue of customers, the number of busy servers, $N(t) + n$, is the same as it would be if there were infinitely many servers. Except for some change from the customary notation, the above formulas are mostly well-known from the theory of the infinite channel service system. The Var $\{N(t) + n\}$ has also been evaluated for the ∞-channel service system. Although rather complicated exact formulas have been derived for this, we wish to use only an approximate formula derived by Haji and Newell[6] under hypotheses equivalent to those described in Section I.3. In the present notation (what was previously called $N(t)$ is now $N(t) + n$) , this formula gives for stationary arrivals $\lambda(t) = \lambda$,

$$I^* \equiv \frac{\text{Var}\{N(t)\}}{E\{N(t)\} + n} \simeq 1 + (I - 1) \frac{\int_0^\infty d\tau \overline{G}^2(\tau)}{\int_0^\infty d\tau \overline{G}(\tau)} \tag{2.21}$$

$$= I + (1 - I) \frac{\int_0^\infty d\tau G(\tau)\overline{G}(\tau)}{\int_0^\infty d\tau \overline{G}(\tau)} \,. \tag{2.21a}$$

One can also derive a more general formula for non-stationary arrivals

$$I^*(t) \;=\; 1 + (I - 1) \, \frac{\displaystyle\int_0^\infty d\tau \, \overline{G}^2(\tau) \lambda(t - \tau)}{\displaystyle\int_0^\infty d\tau \, \overline{G}(\tau) \lambda(t - \tau)} \;, \tag{2.21b}$$

but we will usually use (2.20) or (2.20a).

These formulas describe qualitatively some of the peculiar coupling between the fluctuations in the arrivals and the service. If we define

$$S_m = \min [S , S'] \tag{2.22}$$

where S and S' are two independent random variables each with distribution function G , then

$$\int_0^\infty d\tau \, \overline{G}^2(\tau) = E\{S_m\} \tag{2.23}$$

and

$$\int_0^\infty d\tau \, G(\tau)\overline{G}(\tau) = E\{S\} - E\{S_m\} \;. \tag{2.23a}$$

Thus

$$I^* \;=\; 1 + (I - 1) \frac{E\{S_m\}}{E\{S\}} \;=\; I + (1 - I) \frac{(E\{S\} - E\{S_m\})}{E\{S\}} \;. \tag{2.21c}$$

In the special case $S = s$; $E\{S\} = E\{S_m\}$, (2.23a) has its minimum value of 0 , and I^* reduces to I as in (2.2). Also for $S = s$, $I^*(t) = I$ in (2.21b). Generally, (2.23a) represents a qualitative measure of the "width" of the distribution of S and should be interpreted as being comparable with the standard deviation of S , σ_S . If, for example, S is exponentially distributed, (2.23a) has a value $E\{S\}/2$, and $I^* = (1 + I)/2$. If S has approximately a normal distribution (of small variance), (2.23a) has a value of $\sigma_S (2\pi)^{-1/2}$ and

$$I^* \;\simeq\; I + (1 - I) C_S (2\pi)^{-1/2} \;. \tag{2.24}$$

If $I = 1$, as for Poisson arrivals (or for a clustered Poisson arrival process with expected cluster size b feeding a server that serves batches of size b) , then $I^* = 1$, independent of the service distribution. It is also possible, for a sufficiently broad distribution of S , to make (2.23a) arbitrarily close to $E\{S\}$, in which case again $I^* = 1$, independent of the arrival process (i.e., of I) . The latter property is sort of a dual of the first. If we think of servers arriving and being served by customers instead of the reverse, then the arrival process of servers is the superposition of point processes generated by the service completion times of the n servers. If n and σ_S are sufficiently large, this process looks like a Poisson process over times short compared with $E\{S\}$. Since $N(t) + n$ represents the number of servers in service as well as the number of customers in service, the system also looks like a group of "customers" serving a Poisson stream of "servers," for which I^* should be independent of the customers.

If $I > 1$, I^* must have a value between 1 and I. The two types of fluctuations, arrivals and service, tend to counteract each other; the broader the distribution of S , the closer is I^* to its lower limit of 1. If, however, $I < 1$, I^* must still have a value between I and 1 , but the broader the distribution of S, the closer I^* comes to the upper limit of 1.

If $I \ll 1$ and $C_S \ll 1$, I^* has a value comparable with their sum; (2.24) gives $I^* \simeq I + C_S (2\pi)^{-1/2}$. This would be the typical situation for a bus route with many large vehicles, $b \gg 1$, but also a well controlled trip time $C_S \ll 1$. In this case, the value of I^* will be dominated by fluctuations in customer arrivals or trip time according to which of I or C_S is, in order of magnitude, the larger. If, however, I^* is so small that $\sigma_N \lesssim 1$, the continuum approximations used here do not make much sense. This clearly is what happens if, for buses, the trip time is so well controlled that the buses do not pass each other very often. In such a case one obviously would analyse the problem differently, by looking at the trips of individual vehicles.

The value of I^* will be important not only as a measure of when queueing starts but also how it behaves. Our conclusion here is that, if $\lambda(t)$ is nearly constant, or varies linearly with time over times of the order of the service time, $N(t)$ has an expectation given approximately by (2.18) and a variance given by (2.21), provided that σ_N is at least 2 or 3 and provided $E\{N(t)\}$ plus about $2\sigma_N$ is still negative, i.e.

$$1 - \rho(t - E\{S_0\}) \gtrsim 2\sigma_N/n \simeq 2(I^*/n)^{1/2} . \tag{2.25}$$

If $\lambda(t)$ is increasing with time, we might also expect these formulas to apply at least for some fraction of a mean service time after (2.25) fails, even though $E\{N(t)\}$ may become positive, because the consequences of queueing will not be felt until they distort the rate of service completions and thus the supply of available servers. One also expects the distribution of $N(t)$ to be approximately normal until such time.

3. <u>Approximations with queueing and large $S_k = s$</u> . It is obviously not possible to present and analyse examples of all the peculiar types of queue behaviors that can be generated from curves $A_c(t)$ having many surges (deterministic or stochastic) in arrivals within a time of order $E\{S\}$. We will limit the discussion here to some curves $A_c(t)$ of the general shape shown in Fig. I.4 involving a single rush hour. In this section we will be concerned with the queue behavior particularly in situations where $\rho(t)$ exceeds 1. Again we begin the special case in which $S_k = s$.

One can see from Fig. I.4 that the qualitative shape of $A_s(t)$ or $D(t)$ after queueing starts, and consequently the behavior of the queue size $N(t)$, will depend upon the relative size of s and the duration of queueing, $t_1 - t_0$ in Fig. I.4. Fig. I.4 shows a discontinuity in the slope of $D(t)$ at t_0 where it switches from $A_c(t)$ to $A_s(t)$. Since all customers have the same service time, the rate at which customers enter the service at time t , when there is a queue, is the same as the rate at which servers become free, which, in turn, is the rate at which customers entered the service at time $t - s$. Any discontinuity in the slope of $D(t)$ will be repeated a time s later, if there is still a queue. There are extreme situations in which these irregularities in $D(t)$ are dominant, and others in which they are insignificant.

If we are concerned with people queueing for taxis or buses, airplanes queueing for gate positions, etc., the service time s is likely to be comparable with the duration of the rush hour. Fig. II.1 shows an illustration of a case in which queueing starts at time t_0 , prior to which customers were entering service at a

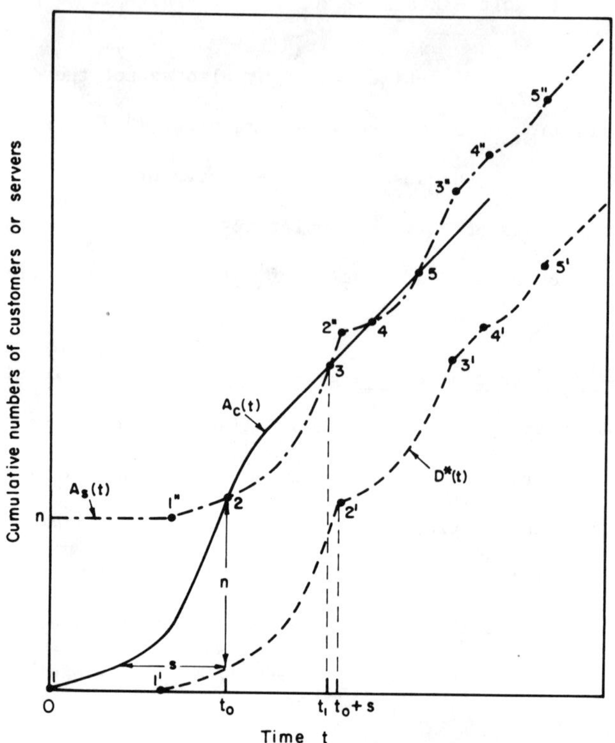

Fig. II.1 - An example of a construction of cumulative curves for which the service rate has a sharp discontinuity when queueing starts (at point 2) causing a queue to reform (at point 4).

fast rate. After queueing starts, customers can enter service only as fast as servers become free, which is determined by the arrival rate at a time s earlier, when the arrival rate was low.

In Fig. II.1, $A_c(t)$ from points 1 to 2 generates the curve 1'2' of completed services, which generates the curve 1"232" of available servers. The queue forms at 2, when the system runs out of servers, and disappears at 3 when there is a surge of service completions. Between 3 and 4 there is an excess of servers.

Fig. II.1 has been drawn so as to show the possibility of further complications propagated from the disturbance at t_0 , even though $A_c(t)$ is quite smooth. The irregularities at 2' or 2" can cause a queue to reform at 4 . The section of curve 2"4 is a reflection of the slow rate of service completions propagated from the low arrival rate at point 1. The system runs out of servers at 4 but recovers at 5 when the surge of service completions reappears.

The curves $D(t)$ and of service completions $D*(t)$ suffer more discontinui-
ties of slope at 3, 4, and 5 or 3', 4' and 5', respectively. The queue could reform
again in the next server time interval, but each time there is queueing, the curve
$D*(t)$ becomes smoother, provided there are no new surges in $A_c(t)$ to initiate new
complications.

To see how stochastic arrivals influence these solutions, one should first re-
call that, although there is a discontinuity in the slope of $D(t)$ at t_0 where the
queue suddenly forms, there is no discontinuity in the number of customers in the
system at t_0 ;

$$N(t) = A_c(t) - A_s(t) = A_c(t) - A_c(t - s) + n \qquad (3.1)$$

is valid until time $t_0 + s$. If $A_c(t)$ is stochastic but s is constant, (3.1)
will be true if $t < t_0 + s$ for all possible values of the random time t_0 . Until
such time, (2.3) or (2.6) also hold.

Another possible interpretation of this is the following. Since in analysing
the number of customers in the system $N(t) + n$, it makes no difference whether a
customer is in the service or in the queue, one can imagine an equivalent hypothetical system in which a customer, upon joining the queue, is assigned a total time
in the system, waiting time plus service time. It makes no difference either, in
which order he spends this time. If he were served first and waited later, it is obvious that $N(t) + n$ would not see the effects of queueing until a time s after
it started, when customers were staying for a time longer than s .

Although $E\{N(t)\}$ can be evaluated correctly from a graph of only $E\{A_c(t)\}$
for $t < t_0 + s$, the same is not true of the queues $N_c(t)$ and $N_s(t)$ of customers and servers, respectively, because

$$N_c(t) \;=\; \max\,[0\,,\,N(t)]$$
$$\qquad\qquad\qquad\qquad (3.2)$$
$$N_s(t) \;=\; \max\,[0\,,\,-N(t)]$$

implies

$$E\{N_c(t)\} \geq \max [0, \quad E\{N(t)\}]$$
$$E\{N_s(t)\} \geq \max [0, - E\{N(t)\}] .$$

$$(3.3)$$

Thus, a deterministic approximation in which $A_c(t)$ is simply replaced by $E\{A_c(t)\}$ will generally lead to an underestimation of the expected queue of both customers and servers, at least over this time range prior to $t_0 + s$.

Actually, much of the stochastic theory of queues can, in a sense, be considered as a study of certain consequences of the fact that the expectation of the maximum of two random variables is larger than the maximum of their expectations. Queues cannot be negative, so an accidental positive queue cannot be compensated by a negative one. Any servers that are accidentally idle cannot recover the lost time by serving someone who has not yet arrived.

Although in (I 2.10) we found it convenient to introduce a single function $N(t)$ to replace the two functions $N_c(t)$ and $N_s(t)$, one of which was zero at each t , if we wish to retain the geometric representation for $E\{N_c(t)\}$ and $E\{N_s(t)\}$, we should reintroduce the curve $D(t)$, the role of which was virtually eliminated by the $N(t)$. Since from section I 2, we had

$$N_c(t) = A_c(t) - D(t) , \quad N_s(t) = A_s(t) - D(t)$$

$$(3.4)$$

and

$$D(t) = \min [A_c(t) , A_s(t)] ,$$

it follows that

$$E\{N_c(t)\} = E\{A_c(t)\} - E\{D(t)\}$$
$$E\{N_s(t)\} = E\{A_s(t)\} - E\{D(t)\} .$$

$$(3.5)$$

Thus expected queue lengths could be evaluated from graphs of $E\{A_c(t)\}$, $E\{A_s(t)\}$, and $E\{D(t)\}$, but

$$E\{D(t)\} = E\{\min [A_c(t) , A_s(t)]\} \leq \min [E\{A_c(t)\}, E\{A_s(t)\}]$$

$$(3.6)$$

can no longer be inferred from $E\{A_c(t)\}$ and $E\{A_s(t)\}$ alone.

The two sides of (3.3) and (3.6) will differ significantly only if the probability distribution of $N(t)$ overlaps 0, or, equivalently, the distributions of $A_c(t)$ overlaps that of $A_s(t)$. This will happen only if $E\{N(t)\} = E\{A_c(t) - A_s(t)\}$ is within one or two standard deviations of zero. Even then, the difference between the two sides of (3.3) or (3.6) will be at most of the order of magnitude of this standard deviation.

In the present situation, not only is (3.1) true until a time s after queueing starts, but it was also postulated in the last section that $N(t)$ should be approximately normally distributed with a variance σ_N^2. If this is true then

$$E\{N_c(t)\} \simeq \int_0^\infty dx \ x(2\pi)^{-1/2}\sigma_N^{-1} \exp \left(- [x - E\{N(t)\}]^2/2\sigma_N^2\right)$$

$$= \sigma_N H(E\{N(t)\}/\sigma_N) \tag{3.7}$$

where

$$H(z) \equiv (2\pi)^{-1/2} \exp (-z^2/2) + z\Phi(z) . \tag{3.8}$$

Fig. II.2 shows a graph of the function $H(z)$. From (3.7) and Fig. II.2 we see that if $E\{N(t)\}$ is negative by even one standard deviation $(z = -1)$,

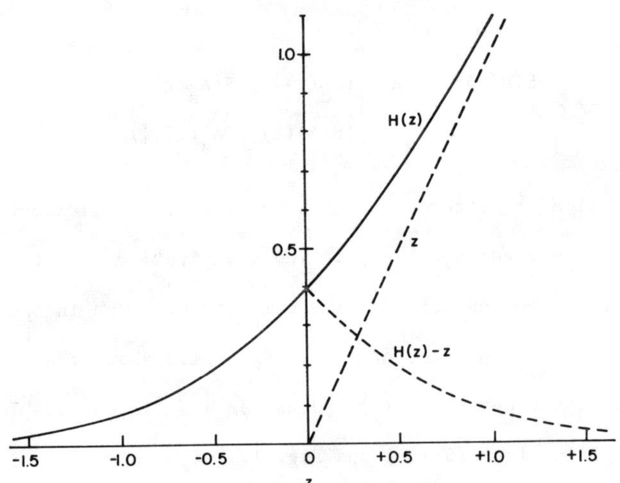

Fig. II.2 - Graph of the functions $H(z)$ (solid line), z, and $H(z) - z$ (broken line) of equation (3.8).

then $E\{N_c(t)\}$ is less than $\sigma_N/10$. This would already be quite small in any of the intended applications. For example, if $I = 1$, $n = 16$, $\sigma_N \simeq 4$, this gives a mean customer queue of about 0.4 as compared with an expected number of about 12 customers in service. Even when $E\{N(t)\} = 0$, $E\{N_c(t)\}$ is only about $0.4\sigma_N$, which, in this example, is about 1.6 as compared with 16 in service.

For $z > 0$, we see that $H(z)$ approaches z quite rapidly for $z \gtrsim 1$. Also for $z > 0$

$$H(z) = z + H(- z) ,$$

so we can write,

$$H(z) = \max [0 , z] + H(- |z|) , \tag{3.9}$$

i.e. the difference between $H(z)$ and $\max [0 , z]$ is symmetric in z , as shown in Fig. II.2.

The formulas (3.3) and (3.6) can thus be replaced by

$$E\{N_c(t)\} \simeq \max [0 , E\{N(t)\}] + \sigma_N H(- |E\{N(t)\}|/\sigma_N) \tag{3.10a}$$

$$E\{N_s(t)\} \simeq \max [0 , - E\{N(t)\}] + \sigma_N H(- |E\{N(t)\}|/\sigma_N) \tag{3.10b}$$

and

$$\begin{aligned} E\{D(t)\} \simeq &\min [E\{A_c(t)\} , E\{A_s(t)\}] \\ &- \sigma_N H(- |E\{A_c(t) - A_s(t)\}|/\sigma_N) , \end{aligned} \tag{3.10c}$$

at least until a time s after queueing starts, i.e., until about a time s after $E\{N(t)\}$ increases to a value of $-\sigma_N$. The quantitative effect of the second terms in (3.10) is that it smooths out the discontinuity of slope in the first term, on a scale of counts measured in units of σ_N . Fig. II.3 shows the same curves as in Fig. II.1, but with the curve $E\{D(t)\}$ drawn as a solid curve. In the vicinity of point 2, this is drawn from (3.10c) and Fig. I.1 with $n = 15$ and $\sigma_N \simeq 4$.

The above formulas which determine $E\{D(t)\}$ until a time s after queueing starts, will determine the expected number of service completions $E\{D^*(t)\} = E\{D(t - s)\}$ and of available servers $E\{A_s(t)\} = n + E\{D(t - s)\}$ until a time $2s$ after queueing starts. Thus, in Fig. II.3, the curve $E\{D^*(t)\}$, solid line, near

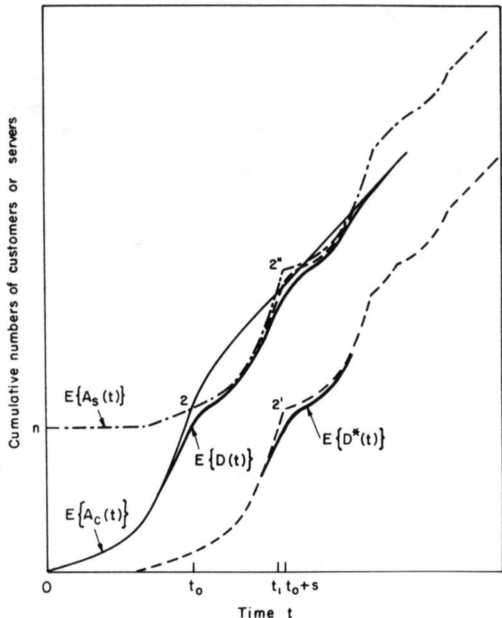

Fig. II.3 - The cumulative curve $E\{A_c(t)\}$ is the
same as the curve $A_c(t)$ of Fig. 1. The graph
illustrates the stochastic correction to $E\{D(t)\}$,
solid line, which rounds the corner at point 2.
This generates also a correction to $E\{D^*(t)\}$,
solid line, at point 2' and to $E\{A_s(t)\}$, broken
line, at point 2". The solid curve $E\{D(t)\}$ near
2" involves a second stochastic correction.

point 2' is a horizontal translation of $E\{D(t)\}$, and $E\{A_s(t)\}$, broken line, near

2" is a vertical translation of this. There are many possible shapes for $E\{D(t)\}$,

however, and Fig. II.3 shows only one (fairly complex) example.

We see from Fig. II.3 that another characteristic time is entering the problem,

the length of time over which $|E\{N(t)\}| \lesssim \sigma_N$. The value of $E\{N(t)\}$ could rise to

a maximum value in the range between $-\sigma_N$ and $+\sigma_N$, and then drop below $-\sigma_N$ again,

or it could pass through this range and attain a maximum value larger than $+\sigma_N$. In

the latter case $E\{N(t)\}$ must eventually come back again passing through this range a

second time with a second (possibly different) characteristic transition time.

In this section we are concerned with "large S" which has been interpreted, so

far, to mean a value comparable with the duration of the rush hour. Actually the

value of s compared with the duration of the rush hour is not particularly critical

to the method of analysis; the more important comparison is between s and the dura-

tion of these transitions. There will be complications if the distortion in $E\{D(t)\}$

due to the second term of (3.10) persists for a time larger than s so as to inter-act with the distortion it caused in $E\{A_s(t)\}$. The extreme case (to be considered later) would be that in which s is small compared with the time when $\left|E\{N(t)\}\right| \approx \sigma_N$ where these stochastic effects may accumulate, a distortion of $E\{D(t)\}$ causes one in $E\{A_s(t)\}$ which further distorts $E\{D(t)\}$, etc.

Consider first the case in which $E\{N(t)\}$ rises to a value less than $+\sigma_N$ and remains between $-\sigma_N$ and $+\sigma_N$ only for a time less than s . In this situation, some realizations $A_c(t)$ cause a customer queue to form, some do not and, in any case, the queue is at most of order σ_N . The curves $E\{A_c(t)\}$ and $E\{A_s(t)\}$ come within σ_N of each other, maybe crossing, maybe not. In either case, $E\{D(t)\}$ is given by (3.10) and lies below both curves.

Fig. II.4 illustrates this. In this figure $E\{A_s(t)\}$ drops slightly below $E\{A_c(t)\}$ between points 2 and 3. The stochastic effects give a smooth curve for $E\{D(t)\}$, cause a slight shift in $E\{D*(t)\}$ near 2'3' and in $E\{A_s(t)\}$ near 2", 3", but the effects propagate no further and leave no residual effects.

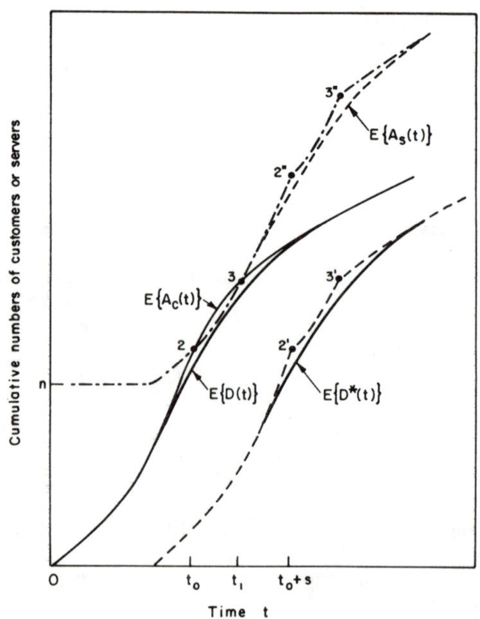

Fig. II.4 - The same type of construction as in Fig. II.3 illustrates a brief period of queueing between points 2 and 3. The effects of queueing disappear after one service time.

One can think of the stochastic effects as being caused by some servers being accidentally late to meet their customers (or customers late to meet their servers). Late or not, however, the customer will be served. A late server will be late to complete service and late to become available again, but if, at this time, there is already a queue of servers, the final count of servers will return to normal once the late server has joined the queue. The effects of lateness will persist only if a server is not available when needed.

Consider next the case in which $E\{N(t)\}$ passes through the range $(-\sigma_N , +\sigma_N)$ within a time less than s . This will create the smooth transition in $E\{D(t)\}$ shown in Fig. II.3 near point 2, and its image at 2' and 2". The first thing to observe about this is that $E\{D(t)\}$ does join the curve $E\{A_s(t)\}$; the accuracy of the deterministic approximations is recovered, at least temporarily. Once it is virtually certain that a queue of customers exists, it is also virtually certain that the expected number of customers to enter service is equal to the expected number of available servers.

If $E\{N(t)\}$ should pass back through zero again, still within a time s after the queueing started, (3.10) would apply over both transitions. If we had drawn Fig. II.4 using a smaller σ_N , the curve for $E\{D(t)\}$ would have tried to follow min $[E\{A_c(t)\} , E\{A_s(t)\}]$ closer than shown. It would just round the corners at 2 and at 3, but, if $E\{N(t)\}$ becomes larger than about σ_N in between, $E\{D(t)\}$ would come in tight to the curve $E\{A_s(t)\}$.

If $E\{N(t)\}$ remains greater than σ_N for several service times, as in Fig. II.5 or I.4, then the construction of $E\{D(t)\}$ will proceed exactly as in Fig. I.4 until $E\{N(t)\}$ comes back down. The curve $E\{D(t)\}$ will follow the equation.

$$E\{D(t - s)\} = n + E\{D(t)\} , \qquad (3.11)$$

starting, however, with the values of $E\{D(t)\}$ evaluated from (3.10) over the first service interval after queueing started. The resulting curve of $E\{D(t)\}$ will differ from that of Fig. I.4 only in that all corners of $E\{D(t)\}$ are smoothed out from below in the same way as the first one. Fig. II.5 shows an example with sharper corners, like Figs. II.1 and II.3 except that the queue last longer.

In the vicinity of point 10 of Fig. I.4, $E\{N(t)\}$ will make another transition

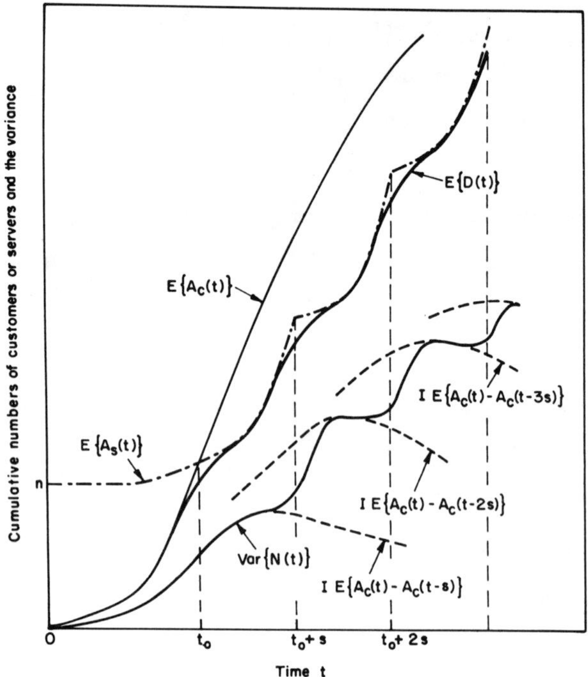

Fig. II.5 — The curves $E\{A_c(t)\}$ and $E\{A_s(t)\}$ are drawn as in Fig. II.3, but the queueing lasts many service times. The rounding of the corners of $E\{D(t)\}$ is repeated at times $t_0 + js$, $j = 1, 2, \ldots$. The curve of $E\{D*(t)\}$ is not shown. The curves Var $\{N(t)\}$ show the sudden rises near $t_0 + js$ between curves of $I\{A_c(t) - A_c(t - js)\}$. The scale of Var $\{N(t)\}$ corresponds to $I = 1/2$.

through 0 and another correction to $E\{D(t)\}$ will be necessary. If $N(t)$ is normally distributed, (3.10c) will still apply with σ_N interpreted as the standard deviation of $N(t) = A_c(t) - A_s(t)$, but it will not have the same value (2.2) as during the first service time after queueing started.

Until time s after queueing started, $N(t) + n$ was the number of arrivals during a service time; it was assumed to be normally distributed with a variance σ_N^2 of approximately $I[N(t) + n]$. For later times, one can use (I 2.11) to obtain the distribution of $N(t)$ from the distribution of $N(t - s)$. Since the arrivals during time $(t - s , t)$ should be approximately normal and independent of $N(t-s)$, the distribution of $N(t)$ will be the convolution of a normal distribution for $A_c(t) - A_c(t - s) - n$ with the distribution of max $[0 , N(t - s)]$, the queue length at time $t - s$, therefore

$$\text{Var}\{N(t)\} \;=\; \text{Var}\{A_c(t) - A_c(t - s)\} + \text{Var}\{\max[0, N(t - s)]\}$$
$$\simeq\; \text{IE}\{A_c(t) - A_c(t - s)\} + \text{Var}\{\max[0, N(t - s)]\} \tag{3.12}$$

Before $N(t - s)$ reaches a value of about $-[\text{Var}\{N(t - s)\}]^{1/2} \simeq -(\text{In})^{1/2}$, at a time t about one service time after queueing starts, the second term of (3.12) was neglected and the first term gave the above value of $I[N(t) + n]$. As $E\{N(t - s)\}$ passes through the range from about $-(\text{In})^{1/2}$ to $+(\text{In})^{1/2}$, the second term of (3.12) increases from about 0 to about $\text{Var}\{N(t - s)\}$, after which

$$\text{Var}\{N(t)\} \;\simeq\; \text{IE}\{A_c(t) - A_c(t - s)\} + \text{Var}\{N(t - s)\} \;.$$

If we resubstitute (3.12) for $\text{Var}\{N(t - s)\}$, this becomes

$$\text{Var}\{N(t)\} \;\simeq\; \text{IE}\{A_c(t) - A_c(t - 2s)\} + \text{Var}\{\max[0, N(t - 2s)]\} \;, \tag{3.13}$$

which has the same form as (3.12) except that the first term is the number of arrivals in a time $2s$, and the last term is evaluated at time $t - s$. The second term of (3.13) is small until t is about two service times after queueing started.

If, during the translation between (3.12) and (3.13) at about one service time after queueing starts, when $|E\{N(t - s)\}| \lesssim (\text{In})^{1/2}$, we can assume that $N(t - s)$ is approximately normally distributed, (3.12) can be approximated by

$$\text{Var}\{N(t)\} \;\simeq\; \text{IE}\{A_c(t) - A_c(t - s)\}$$
$$+\; \text{Var}\{N(t - s)\} H^*(E\{N(t - s)\}/(\text{In})^{1/2}) \tag{3.14}$$

where

$$H^*(z) \;=\; (2\pi)^{-1/2} \int_0^\infty dx\, x^2 \exp[-(z - x)^2/2] - H^2/z)$$
$$\tag{3.15}$$
$$=\; (2\pi)^{-1/2}\, z \exp(-z^2/2) + (1 + z^2)\Phi(z) - H^2(z) \;.$$

The function $H^*(z)$ is shown in Fig. II.6. It goes from a small value for $z \lesssim -1$, to nearly 1 at $z \gtrsim +1$, giving a smooth transition between (3.12) and (3.13), but on the same time scale as the correction to $E\{D(t)\}$ in Fig. II.3 .

When $E\{N(t - 2s)\}$ in (3.13) passes through the range $-(\text{In})^{1/2}$ to $+(\text{In})^{1/2}$

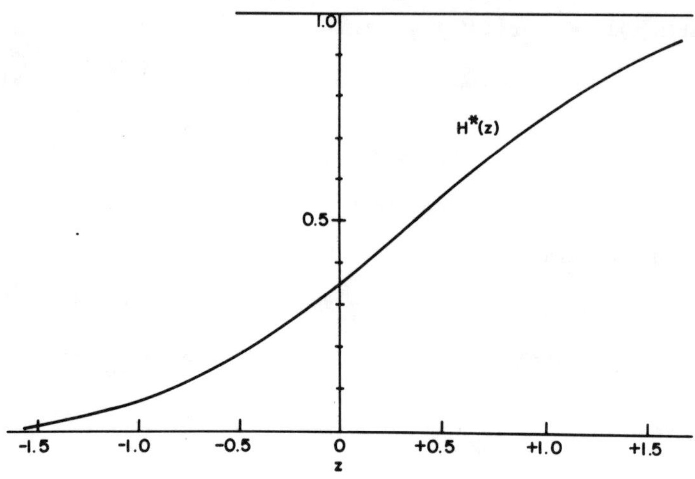

Fig. II.6 - The function H*(z) of
equation (3.15).

at about two service times after queueing started, the second term of (3.13) now

rises from 0 to Var{N(t - 2s)} . Over this time

$$Var\{N(t)\} \simeq IE\{A_c(t) - A_c(t - 2s)\}$$
$$+ Var\{N(t - 2s)\}H*(E\{N(t - 2s)\}/(In)^{1/2}) \tag{3.14a}$$

leading to a value of

$$E\{N(t)\} \simeq IE\{A_c(t) - A_c(t - 3s)\} + Var\{max [0 , N(t - 3s)]\} . \tag{3.13a}$$

These differ from (3.14) and (3.13) only in that t - js , j = 1, 2, ... has been

replaced by t - (j + 1)s . This pattern continues as long as the queueing persists.

If queueing continues for many multiples of the time s , Var{N(t)} will become

approximately I times the expected number of arrivals since queueing started. This

variance does not increase smoothly, however. It rises rather sharply at times which

are integer multiples of time s after queueing starts, i.e., at the same times when

$E\{D(t)\}$ is rounding the corners of Fig. II.5.

Fig. II.5 shows Var{N(t)} on the same graph with $E\{A_c(t)\}$. The value of n

is a scale parameter for $E\{A_c(t)\}$ and I a scale factor for Var{N(t)} . The

curves are actually drawn with n = 15 , and I = 1/2 but we are more concerned with

the shape than the scale. The curve for Var{N(t)} rises smoothly (first term of

(3.13)) until a time almost $t_0 + s$, a time s after queueing starts. It then takes a sharp rise (following (3.14)) to a new plateau; rises again near time $t_0 + 2s$ (following (3.14a)), etc., almost like a step function.

During the sudden rises in $\text{Var}\{N(t)\}$, the distribution of $N(t)$ is approximately the convolution of a normal distribution with a truncated normal distribution originating from the queue distribution when queueing started. Between these rises, however, $N(t)$ is approximately normal. It will also be approximately normal during the rises if queueing persists long enough so that the number of arrivals since the start of queueing becomes much larger than n . The argument here is that if one takes the sum of a normal random variable (the arrivals) and another independent random variable (the queue) of relatively small variance, the sum will also be approximately normal.

The value of $E\{D(t)\}$ near point 10 of Fig. I.4 or near point 3 of Fig. II.1 can be evaluated explicitly from the value of $E\{\max [0 , N(t)]\}$ even if the distribution of $N(t)$ is itself a convolution of a normal and a truncated normal distribution. The details of this, however, are somewhat involved and do not seem to be of great practical importance. The peculiar manner of growth of $\text{Var}\{N(t)\}$ described above is characteristic of a system with a fixed service time $S_k = s$, because the service, while busy, has a perfect memory of its past stochastic behavior. This will not be true for a system with random S . Qualitatively, one will find that, unless point 3 of Fig. I.3 lies near a corner of $E\{A_s(t)\}$, integer multiples of time s after queueing starts, $N(t)$ will be approximately normal and (3.10c) will describe the transition accurately enough. The appropriate value of σ_N in (3.10c) is, however, likely to be considerably larger than at the start of queueing. On the other hand, if this second transition occurs at the end of the rush hour, it is also likely that $E\{N(t)\}$ is decreasing so rapidly that the transition will not last very long.

Figs. IL.1 or II.3 show about as complicated an example as one is likely to encounter with a single rush hour. The queue, in the deterministic approximation, tries to disappear near point 2" which is close to a service time after queueing starts; it reforms again at point 4 and finally disappears at point 5. This will happen for some, but not all realizations. Because of stochastic variations near point 2, there is a net average loss in the number of available servers, i.e., a drop in $E\{A_s(t)\}$, near point 2'. If this does not cause $E\{N(t)\}$ to become positive near 2", it may at least cause $N(t)$ to be positive for some realizations, i.e., $E\{N_c(t)\} > 0$, in which case $E\{N_c(t)\}$ dips to a low value near point 2" but does not vanish. The solid line of Fig. II.3 near point 2" shows a likely behavior of $E\{D(t)\}$. The details of this and other possible behaviors are straightforward but somewhat tedious to analyse. We choose not to pursue this further.

Before we go on to the case of random service time, it should perhaps be emphasized again, that for $S = s$, the graphical construction in which one simply replaces $A_c(t)$ by $E\{A_c(t)\}$ is indeed a very useful method for estimating $E\{N_c(t)\}$, $E\{N_s(t)\}$, etc. They are typically in error only by about $1/2 \; (nI)^{1/2}$ at most, and then only at certain critical times. This is to be compared with a typical value of n customers in the service. Although it is helpful to know the order of magnitude of errors, it seems unlikely, in the situations described in this section, that one would need to know the queue lengths to such an accuracy that one need evaluate the errors carefully and correct for them. The one exception would be that in which one has designed a system to have small queues and the queues which do exist are due almost entirely to the stochastic effects.

These deterministic approximations always underestimate $E\{N_c(t)\}$ and overestimate $E\{D(t)\}$, because an accidental loss in service cannot be compensated by an accidental excess of servers. The approximations also underestimated $E\{N_s(t)\}$ when the queueing of customers first started, but this will not necessarily be true at later times. Although there are certain similarities in the treatment of servers and customers, they do not have equivalent properties. The curve $A_c(t)$ and its stochastic properties were considered as given, whereas $A_s(t)$ is derived from it. An accidental excess of servers at one time will generate a deficiency at a later time (about time s later) because the total number of servers, n , is fixed.

4. <u>Queueing with random S , $C_s \ll 1$</u>. Many of the properties described above for $S = s$ will also be true in the case of random service times. Random service times are certainly more difficult to analyse, either analytically or graphically, and they introduce some new effects, but, on the other hand, some of the unpleasant

consequences of the long time memory of the service will be diminished. Also, as we

saw in Section 2, fluctuations in the service may, under some conditions, compensate

for the fluctuations in the arrivals.

Since $S = s$ is a special case of random service with $\sigma_S^2 = \text{Var} \{S\} = 0$, one

would expect the behavior of a system with $C_S \ll 1$ to be similar to that of

$C_S = 0$. In Section 2, we considered the behavior of $N(t)$ prior to the time

queueing starts. Over this time, $E\{A_s(t)\}$ can be evaluated from (2.11) with

$E\{D(t - \tau)\}$ replaced by $E\{A_c(t - \tau)\}$. The main effect here of a small variance

in S is to smooth out any irregularities that might exist in $E\{A_c(t)\}$, particu-

larly those with a time scale of order σ_S . If, for example, some component of the

arrival stream has a predictable surge (some customers arrive at nearly the same

time each day), a constant service time s would cause a predictable surge in $A_s(t)$

at a time s later. If there were no queue of servers at the later time, so that

the returning servers were put into immediate service again, then this surge would

propagate further in time. This is, of course, the cause of the repeated irregu-

larities in $E\{D(t)\}$ described in the last section. The effects of any distur-

bances would, with random S , be smeared out over a time of order σ_S .

The service acts on realization $A_c(t)$, not $E\{A_c(t)\}$, and does not know

whether a surge in $A_c(t)$ is deterministic or stochastic until it has served other

realizations and taken the average. If there is a random surge in $A_c(t)$, the ser-

vice will still tend to smear its effects. This is, in essence, the explanation of

why, for $I > 1$, Var $\{N(t)\}$ in (2.21) is less for random S than for $S = s$.

Consider now what will happen in some of the examples treated in the last sec-

tion if S were random with $C_S \ll 1$. Until queueing starts near point 2 of Figs.

II.1 or II.3, $E\{D*(t)\}$ and $E\{A_s(t)\}$ will remain nearly as shown, but perhaps be

smoothed slightly if $E\{A_c(t)\}$ fluctuates appreciably over a time of order σ_S .

As we approach point 2, $N(t)$ will still have approximately a normal distribution

but with a variance given by (2.21), slightly higher than for $S = s$ if $I < 1$,

smaller if $I > 1$.

If S has a narrow distribution (width comparable with σ_S) , then any server

that was used at time t will not likely be available again until at least time

$t + E\{S\} - \sigma_S$. The number of servers used by time t will, therefore, determine $\Lambda_s(\tau)$ until about time $\tau \simeq t + E\{S\} - \sigma_S$. If at time t there is no queueing, $N(t) < 0$, so that $D(t) = A_c(t)$, then the properties of $N(\tau)$ or $N(\tau) + n$ described in Section 2 will still be true until about time $\tau \simeq t + E\{S\} - \sigma_S$, thus until a time about $E\{S\} - \sigma_S$ after queueing first starts; in particular $N(\tau)$ should be approximately normally distributed with a variance of σ_N^2 of approximately $I*[N(t) + n]$.

The shape of $E\{D(t)\}$ near point 2 of Figs. II.1 or II.3 remains essentially unchanged from that described in the last section for $\sigma_S = 0$ except for a slight change in the value of σ_N used in (3.10). The width and shape of the transition at point 2 is rather insensitive to σ_S but is determined mainly by the shape of $E\{A_c(t)\}$ and the value of σ_N .

We have now created a rather sharp transition in $E\{D(t)\}$, the width of which has been assumed to be small compared with $E\{S\}$, but essentially unrelated to the value of σ_S . The subsequent behavior of the curve for $E\{A_s(t)\} = n + E\{D*(t)\}$ is to be evaluated now from (2.11) and $E\{D(t)\}$. Regardless of how sharp the transition in $E\{D(t)\}$ near point 2 of Fig. II.3, the transition for $E\{D*(t)\}$ and $E\{A_s(t)\}$ at points 2' and 2", respectively, must have a width of at least σ_S , and, in any case, be a smoothing on a scale of σ_S of the behavior shown in Fig. II.3.

If queueing persists for many service times as in Fig. II.5 or I.4, the stochastic behavior of $A_s(t)$ or $D*(t)$ after queueing becomes certain, is independent of any new arrivals. Servers are put back in service as soon as they become free. In Fig. I.4, the corner in $D*(t)$ at point 8 will be rounded on a scale of at least σ_S . The next corner in $D*(t)$, at about a time $E\{S\}$ later, will be a further smoothing of a translation of the curve near point 8, etc. Since service times are assumed to be independent of each other, one can also think of the curves $D*(t)$ or $A_s(t)$ as being generated by the superposition of n renewal processes initiated during the first service time after queueing starts.

At least for the first few service times after queueing starts, the mth corner of $D*(t)$ is rounded as if each server served exactly m customers in a new service time equal to the sum of m independent S_k . The variance of the sum of m S_k's is $m\text{Var}\{S\}$, the standard deviation $m^{1/2}\sigma_S$. Thus the mth corner of $E\{D*(t)\}$ is like a smoothing of the first corner on a scale of about $m^{1/2}\sigma_S$.

If queueing persists for such a long time, approximately $mE\{S\}$, that $m^{1/2}\sigma_S$ is comparable with $E\{S\}$, then one can no longer think in terms of "rounding the corners" of $E\{D*(t)\}$ because the rounding of one corner overlaps the next. Eventually, if m becomes so large as to create extensive overlapping, each component renewal process generated by the same server begins to behave like a stationary renewal process, independent of its behavior when queueing started. According to the "fundamental theorem of renewal theory"[3], each server will serve at a constant rate of $1/E\{S\}$ services per unit time. The combined service rate of all servers, the slope of $E\{D*(t)\}$, $E\{D(t)\}$, or $E\{A_S(t)\}$, will become equal to $n/E\{S\}$, (asymptotically) independent of t.

At the moment, this asymptotic behavior for large m is somewhat academic because we have assumed that $E\{S\}$ is comparable with the duration of the rush hour. Yet, to reach an equilibrium behavior, queueing must persist for about m service times with $m^{1/2}\sigma_S \gg E\{S\}$, i.e. for about $m \gg C_S^{-2}$ service times. We have further assumed that $C_S \ll 1$. The above equilibrium behavior, however, does give an indication of what to expect when we consider smaller values of $E\{S\}$ and/or larger values of C_S.

During the time when there is a queue of customers, $\text{Var}\{N(t)\}$ will be increasing. This will eventually influence the properties of $N(t)$ when it goes through the transition back to the state of no-queueing. For $\sigma_S = 0$, we saw in Section 3 that $\text{Var}\{N(t)\}$ grew approximately like the variance of the number of arrivals since queueing started, except that the growth was almost like a step function with sudden rises near multiples of a service time after queueing starts. For sufficiently small σ_S, this same pattern must still exist. The effect of the σ_S is to smooth these stops on a scale of order $m^{1/2}\sigma_S$ at the mth step, much as it smoothed the corners of $E\{D*(t)\}$, $E\{D(t)\}$, etc.

This smoothing of the step function will cause a larger variance than for $\sigma_S = 0$ on the early side of the step, but a smaller variance than for $\sigma_S = 0$ on the late side of the step. The latter effect is again analogous to the competition between fluctuations in servers and arrivals that exist for $I > 1$. The first transition, when queueing starts, introduces a large variance in $N(t)$ over a short time, much as would be created by a batch of customers arriving at a predictable time but with a random batch size. The fluctuations in the service time tend to spread the after-effects of this disturbance over time and thereby reduce the intensity of its effects.

Except at times where Var $\{N(t)\}$ is influenced by this smoothing, a small σ_S will have little effect on Var $\{N(t)\}$. Although σ_S will cause fluctuations in the time at which a particular server starts service (this is what smooths the steps), there will be little fluctuation in the count of servers at intermediate times.

In some suitable time average sense, the σ_S will, however, have a net effect of increasing Var $\{N(t)\}$. The $A_c(t)$ and $D(t)$ are statistically dependent because $A_c(t)$ influences the starting times of servers when queueing first starts, in particular, it introduces statistical dependences between the starting times of the n servers. Subsequently, however, the service times of all servers and the number of new arrivals are statistically independent.

Eventually, if queueing persists long enough so that $m^{1/2}\sigma_S$ becomes comparable with $E\{S\}$ and the smoothing of the steps in Var $\{N(t)\}$ begin to overlap, these statistical dependencies caused by the starting times will decay. The properties of the servers will become statistically independent of each other and of $A_c(t)$ as the renewal processes approach a stationary behavior that is independent of its starting time.

For a single renewal process of service time S , the variance of the number of services (renewals) is, asymptotically, approximately C_S^2 times the mean number of services[3]. For the superposition of n independent renewal processes both the mean and the variance of the number of services are n times those of a single process. Thus the same relation is also true of the superposition. If the servers are busy from time τ to time t with

$$t - \tau \simeq mE\{S\} \gg E\{S\}/C_S^2$$

then

$$\text{Var } \{D(t) - D(\tau)\} \simeq C_S^2 E\{D(t) - D(\tau)\} = C_S^2 (t - \tau)n/E\{S\} \ . \qquad (4.1)$$

During this time, the queue increases by $N(t) - N(\tau) = [A_c(t) - A_c(\tau)] - [D(t) - D(\tau)]$. Since the number of arrivals and services are independent, as long as $N(t) > 0$

$$\begin{aligned}
\text{Var } \{N(t) - N(\tau)\} &= \text{Var } \{A_c(t) - A_c(\tau)\} + \text{Var } \{D(t) - D(\tau)\} \\
&\simeq \ \text{IE } \{A_c(t) - A_c(\tau)\} + C_S^2 E \ \{D(t) - D(\tau)\} \ .
\end{aligned}$$

This will hold even if τ is rather close to the time when queueing started and t is rather close to the time when it ends, and the expected number of arrivals during (τ , t) is approximately equal to the expected number of services. The long time effects of a random service time, therefore, is to increase the effective value of I to

$$I^\dagger \equiv I + C_S^2 \qquad (4.2)$$

during times when $N(t) > 0$.

Again this asymptotic behavior is somewhat academic here because it applies only if queueing persists for a time comparable with $E\{S\}/C_S^2$, a much longer time than we expect in the present examples. This will be important later, however, when we consider small values of $E\{S\}$ and/or large C_S . It is also important to note that the effective value of I , $I^\dagger$, for $N(t) > 0$, is not necessarily the same as the effective value of I , I^* , of (2.21), for $N(t) < 0$.

The main purpose here in estimating the variance was to determine an appropriate value σ_N to use in (3.10c) in order to evaluate $E\{D(t)\}$ over the second transition back to $N(t) < 0$. Basically our conclusion is that a small value of σ_S does not change very much the estimates made for $\sigma_S = 0$. Some variance for S may actually help justify disregarding some of the consequences of the fact that Var $\{N(t)\}$ increases in steps for $S = s$. We do not care to pursue this in greater detail here either, because the qualitative estimates made so far would seem to indicate that the quantitative behavior is of little practical concern in any design criteria.

5. <u>Queueing with random S , $C_S \sim 1$</u>. Most of the effects of random service were introduced already in the last section. It remains only to indicate how these effects will influence the behavior of the system when C_S is not small. We assume, however, that C_S is comparable with 1; we do not consider the case of a very broad distribution of S with C_S large compared with 1.

For any service distribution, $E\{A_s(t)\}$ is still related to $E\{D(t - \tau)\}$ through (2.11). This equation was used in Section 2 to determine $E\{A_s(t)\}$ prior to the time that queueing started, during which time $E\{D(t - \tau)\}$ was equal to the given $E\{A_c(t - \tau)\}$. If $\lambda(t)$ were slowly varying, one could obtain simple estimates of $E\{A_s(t)\}$, or equivalently $E\{N(t)\}$ from (2.18), otherwise it might be necessary to evaluate (2.11) by numerical integration. We expect that during this time $N(t)$ will be approximately normally distributed with a variance given by (2.21) or (2.21b).

Once queueing starts, there will again be a range of times, say from τ_0 to τ_1 , during which the distribution of $N(t)$ overlaps 0 and stochastic effects influence the value of $E\{D(t)\}$. In the last section we could evaluate the properties of $N(t)$ until almost a mean service time after queueing started simply by observing that the results of Section 2 were correct at least until queueing delayed the service <u>completion</u> time of some customer. We are still concerned with "large S", which we again interpret to mean that $E\{S\}$ is large compared with the duration $\tau_1 - \tau_0$ of the transition. We will not try to apply the results of Section 2 out to a time of the order of $E\{S\}$ after queueing starts, but we do want to use them over the transition period, until time τ_1 . In particular, for $t < \tau_1$, we wish to estimate $E\{N(t)\}$ or $E\{A_s(t)\}$ from (2.11) or (2.18) with $E\{D(t - \tau)\}$ replaced by $E\{A_c(t - \tau)\}$, and to estimate $E\{N_c(t)\}$ and $E\{D(t)\}$ for $\tau_0 < t < \tau_1$ from (3.10) with σ_N evaluated from (2.21) or (2.21b).

The errors in these approximations are due to the possibility that a queue of customers existed at some time prior to time τ_1 , which, in turn, delays a service completion that would otherwise have occurred also before time τ_1 . But we assumed that there is a negligible probability of a queue prior to time τ_0 , therefore the service must be completed in a time less than $\tau_1 - \tau_0$ in order to affect the results. The order of magnitude of the error in $E\{D(t)\}$, due to the above approximations, can be no larger than the expected queue length at some time before time τ_1, times the expectation that a service is completed in a time of at most $\tau_1 - \tau_0$. Since $E\{N_c(t)\}$ is increasing with time, the error in $E\{D(t)\}$ or $E\{N_c(t)\}$ is, at most, $E\{N_c(t)\}G(\tau_1 - \tau_0)$, but is likely to be only a fraction of this. To justify the approximation, however, it may be necessary to make a somewhat stricter interpretation of "large S," and assume that

$$G(\tau_1 - \tau_0) \ll 1 , \tag{5.1}$$

i.e. there is only a small probability that a service can be completed in the time it takes $N(t)$ to go through the transition.

If we thought it were appropriate, we could evaluate a "second order approximation" in which we estimate the errors in $N(t)$ at time t, $t < \tau_1$, by estimating the extra queue that results at time t due to the possible existence of a queue at earlier times $t - \tau$, but use the above approximations for $N(t - \tau)$. This is a straightforward calculation, but the formulas are somewhat cumbersome and of doubtful practical value. In Chapter III we will, however, be discussing the case of "small S" in which these errors accumulate and may become significant.

For some period of time after time τ_1 (until the queueing ceases), queueing is assumed to be virtually certain, and therefore $D(t) = A_s(t)$; servers are being used as soon as they become free. Equation (2.11) is still valid for $t > \tau_1$, but will now be written in the form

$$E\{D(t)\} = E\{A_s(t)\} = n + \int_0^{t-\tau_1} dG(\tau)E\{A_s(t - \tau)\} \tag{5.2}$$
$$+ \int_{t-\tau_1}^{\infty} dG(\tau)E\{D(t - \tau)\} .$$

The integral over τ has been divided into two parts. In the first part $t - \tau > \tau_1$ and $E\{A_s(t - \tau)\} = E\{D(t - \tau)\}$; the integrand contains the unknown values of $E\{A_s(t - \tau)\}$. In the second part $t - \tau < \tau_1$ and $E\{D(t - \tau)\}$ is considered to have been already evaluated as described above.

Equation (5.2) is an integral equation for the function $E\{A_s(t)\}$ of a type well-known in electric circuit theory, renewal theory, etc. In effect, we have already found solutions of this equation for narrow distributions $G(\tau)$. If $G(\tau)$ is negligible for τ less than about $E\{S\} - \sigma_S > 0$, the first integral of (5.2) vanishes and $E\{A_s(t)\}$ can be evaluated from the known second term for $t \lesssim \tau_1 + E\{S\} - \sigma_S$. Having found $E\{A_s(t)\}$ up to this time, however, we can now include this in the second integral, replacing τ_1 by $\tau_1 + E\{S\} - \sigma_S$. This method is essentially equivalent to that used in Section 4 where we moved forward iteratively in steps of about $E\{S\}$.

For broader distributions of S, the easiest way to solve (5.2) is probably still to move forward iteratively in time, but by smaller increments. For sufficiently small $t - \tau_1$, one may either neglect the first integral of (5.2) or estimate its value from some trial solution. Having evaluated a new estimate of $E\{A_s(t)\}$ from (5.2), one may, if necessary, resubstitute this estimate back into the first integral and recalculate a second estimate of $E\{A_s(t)\}$. Once $E\{A_s(t)\}$ has been calculated accurately over some small time interval, one can shift the value of τ, as above. This may be somewhat tedious (depending upon the accuracy desired) but it is straightforward.

If queueing persists for several service times, and $G(\tau)$ is rather broad, the curve for $E\{A_s(t)\}$ will converge quite rapidly to a straight line asymptote of slope $n/E\{S\}$ as described in Section 4. We will also postpone further consideration of this until Chapter III because this is more relevant to the case of "small S."

Our conclusion here is that, if (5.1) is true, the case $C_S \sim 1$ is qualitatively what would be expected from an extrapolation of the results in Section 4. Only the quantitative details have become more complicated.

CHAPTER III - APPROXIMATIONS FOR SHORT SERVICE TIMES

1. _Introduction._ Chapter II dealt mostly with the behavior of systems for which
the expected service time $E\{S\}$ was comparable with the duration of the rush hour,
but, in any case, $E\{S\}$ was large compared with the time spent going through any
transition period when it was uncertain as to whether there was a queue of customers
or of servers. The analysis was fairly straightforward, though, at times, somewhat
tedious, because we chose to consider a more or less arbitrary curve $E\{A_c(t)\}$ for
the expected number of arriving customers, which varied considerably on a time scale
comparable with the service time. The complications were due mostly to a variety of
possible shapes of the curves $E\{A_c(t)\}$, $E\{D(t)\}$, etc., in the deterministic
approximations. Stochastic effects were relatively small but also easily handled because they did not accumulate.

Here we will consider mostly cases in which $E\{S\}$ is small compared with the
duration of the rush hour. This will, at first, give some simplification because we
will assume that $E\{A_c(t)\}$ is slowly varying on a time scale of order $E\{S\}$; we
can employ approximations such as used in Section II.2, particularly (II 2.17). For
τ of order $E\{S\}$ we may write

$$\lambda(t - \tau) \simeq \lambda(t) - \tau d\lambda(t)/dt , \qquad (1.1)$$

and, for the traffic intensity $\rho(t)$

$$\rho(t - \tau) \simeq \rho(t) - \tau\alpha(t) \qquad (1.2)$$

with

$$\alpha(t) \equiv d\rho(t)/dt = [E\{S\}/n]d\lambda(t)/dt .$$

In certain cases, however, particularly if $\alpha(t)$ vanishes (at the peak of the rush
hour), we may find it necessary to use a quadratic approximation

$$\rho(t - \tau) \simeq \rho(t) - \tau\alpha(t) - (\tau^2/2)\beta(t) \qquad (1.2a)$$

with

$$\beta(t) \equiv - d^2\rho(t)/dt^2 .$$

These formulas will greatly simplify the deterministic approximations and any of the other results of Chapter II, which will still apply if the duration of the transition periods are so short as to be less than $E\{S\}$. If, however, $E\{S\}$ is less than the duration of the transition, we must examine more carefully the possible accumulation of stochastic effects mentioned in Section II.5.

Section 2 describes some of the simplifications that result from applying (1.1) or (1.2) to the formulas of Chapter II. Section 3, as an introduction to the treatment of the transition, deals with stochastic queueing when the queues remain small but queueing may exist for times large compared with $E\{S\}$. Sections 4 and 5 describes some qualitative properties of various types of queue evolutions through the transition.

2. <u>Deterministic approximations</u>. In Section II.2, we have already used (1.1) to describe the behavior of the system when there is no queueing. It remains only to examine the application of (1.1) to the deterministic approximations when the maximum value of $\rho(t)$ exceeds 1. We again, as in Section II.3, consider first the case of constant service time $S = s$.

Fig. III.1 shows a magnified view of a graph similar to Fig. I.4. Starting with only the curve $E\{A_c(t)\}$, it is very easy to construct a straight line of slope n/s which is tangent to $E\{A_c(t)\}$ at the first time, t_0^* , when $\lambda(t_0^*) = n/s$. This is the line

$$E\{A_c(t_0^*)\} + (t - t_0^*)(n/s) \tag{2.1}$$

of Fig. III.1.

If we neglect stochastic effects, queueing will start when $E\{N(t)\} = 0$, or, according to (II 2.6b), at a time t_0 when $\lambda(t_0 - s/2) = n/s$, i.e. at time

$$t_0 = t_0^* + s/2 , \tag{2.2}$$

one-half a service time after the traffic intensity reaches 1. The construction of $E\{D(t)\}$ so that $E\{D(t + s)\} = n + E\{D(t)\}$,(II 3.11), starting from $t = t_0 - s$, guarantees that the line (2.1) is tangent to $E\{D(t)\}$ at each time $t_0^* + ms$, $m = 0 , 1 , \ldots$ until $E\{D(t)\}$ crosses $E\{A_c(t)\}$ again at the termination of the

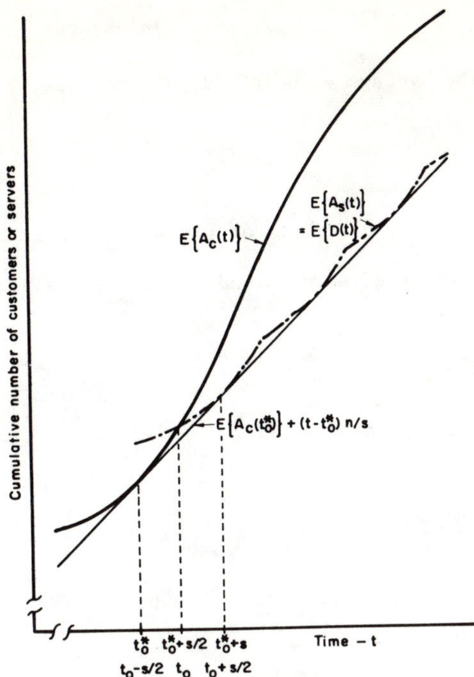

Fig. III.1 – The cumulative arrival of customers and servers for constant service time s . The straight line tangent to $E\{A_c(t)\}$ at t^*_0 is an approximation to $E\{A_s(t)\} = E\{D(t)\}$.

queueing period.

If $\lambda(t)$ changes little during a time s , the line (2.1) is a very good approximation to $E\{D(t)\}$. The difference between the two is periodic in time with a value

$$E\{D(t)\} - [E\{A_c(t^*_0)\} + (t - t^*_0)n/s]$$

$$\simeq \frac{1}{2}(t - t^*_0)^2 d\lambda(t^*_0)/dt^*_0 \quad \text{for} \quad |t - t^*_0| \leq s/2 \ . \tag{2.3}$$

in the period before time t_0 . If we let

$$y = (t - t^*_0)(2/s)$$

so that $|y| \leq 1$, then (2.3) can also be written as

$$y^2(n/8)[s\alpha(t^*_0)] \ . \tag{2.3a}$$

In most practical applications, (2.3a) is probably insignificant. The $s\alpha(t^*_0)$ is the change in traffic intensity during one service time, assumed here to be small compared with 1. In addition to this, one would ordinarily measure counts relative to the number of servers, n , and the factor $y^2 n/8 < (n/8)$ would itself be considered rather small. The average value of (2.3a) over all y is only

$$(n/24)[s\alpha(t_0^*)] . \tag{2.3b}$$

In the absence of stochastic effects at times $t_0 - s/2$ and $t_0 + s/2$, $(t_0^*$ and $t_0^* + s)$, the queues of servers and customers, respectively, are given by

$$- E\{N(t_0 - s/2)\} \simeq E\{N(t_0 + s/2)\} \simeq (n/2)[s\alpha(t_0)] . \tag{2.4}$$

We can apply the stochastic correction (II 3.10) during the time interval $(t_0 - s/2, t_0 + s/2)$ if it is almost certain that $N(t_0 - s/2) < 0$ and $N(t_0 + s/2) > 0$, which will be true for

$$(n/2)[s\alpha(t_0)] \gtrsim \sigma_N \simeq (In)^{1/2}$$

i.e.

$$[s\alpha(t_0)] \gtrsim 2(I/n)^{1/2} . \tag{2.5}$$

We have already assumed that $s\alpha(t_0)$ was "small," but only small enough to justify the use of (1.2) over approximately a range of time $t_0 \pm s/2$. This condition restricts $A_c(t)$ but does not involve n . We have also assumed that $n \gg 1$, so that the lower bound (2.5) on $s\alpha(t_0)$ is "small." For sufficiently large n , there is sizeable range of possible values of $[s\alpha(t_0)]$ sufficiently small to justify (1.2), $(s\alpha(t_0) < 1$ might suffice), but sufficiently large to satisfy (2.5) and thereby guarantee that stochastic effects last less than about one service time. In typical applications such as those described in Section I.1 or II.2, however, the right hand side of (2.5) is not very small. For $I = 1$, $n = 16$, (2.5) would require $s\alpha(t_0) \gtrsim 1/2$, which would mean that $\rho(t)$ goes from a value of about 1 at time $t_0 - s/2$ to at least 1.5 at time $t_0 + s/2$. Under such conditions one would probably not use (1.2) but would go back to the methods described in Part II. In the case of nearly regular arrivals or bulk service (small I), there may be some potential applications, however, in which (2.5) would be satisfied.

The stochastic correction (II 3.10) and the difference (2.3) have opposite signs and therefore tend to cancel. If $s\alpha(t_0)$ were about $(3.2)\,(I/n)^{1/2}$, the

stochastic correction at t_0 (about $0.4\sigma_N$) would just about cancel the difference

(2.3a) at $y = 1$. Although (2.3a) and II 3.10) have quite different analytic forms

over $|y| < 1$, they have very similar shape and would nearly cancel throughout the

whole range $|y| < 1$; (2.1) would be a very accurate approximation to $E\{D(t)\}$

throughout the entire range of times when queueing occurs.

If, however, $s\alpha(t_0) \lesssim (3.2)(I/n)^{1/2}$, which would seem to be the more typical

situation in applications, the stochastic effects would dominate the corrections in

(2.3a) and $E\{D(t)\}$ would lie below the line (2.1). Under these conditions, there

is, at best, only a narrow range of values for $\alpha(t_0)$ before (2.5) fails along with

the estimates in (II 3.10). Stochastic effects start to accumulate. This situation

will be analyzed in the later sections.

Although we have assumed here that $\rho(t)$ exceeds 1 so that queueing becomes

certain, it is conceivable that $\rho(t)$ could satisfy (1.2a), reach a maximum value

with $|N(t)| \lesssim \sigma_N$, particularly $N(t) < -\sigma_N$, and stay in this range for a time

less than s as in Fig. II.4. In this case one need only apply the correction

(II 3.10) to the formulas of Section II.2. We have nothing new to add here.

For random service times, the obvious generalization of (2.1) is the straight

line

$$E\{A_c(t_0^*)\} + (t - t_0^*)(n/E\{S\}) \tag{2.6}$$

of slope $n/E\{S\}$, which is tangent to $E\{A_c(t)\}$ at time t_0^* when $\lambda(t_0^*) = n/E\{S\}$.

If we neglected stochastic queueing effects over the transition, then, according to (II 2.18), (II 2.18a), queueing would start at time

$$t_0 = t_0^* + \frac{1}{2}\, E\{S\}(1 + c_S^2) = t_0^* + E\{S_0\} \ . \tag{2.7}$$

If, in making comparisons with $S = s$, we choose the same curve $E\{(A_c(t))\}$ and

$E\{S\} = s$, then t_0^* is independent of C_S and t_0 increases with C_S . The first

effect of $C_S > 0$ is to delay the start of queueing.

For $t_0^* < t < t_0$, the difference between $E\{D(t)\} = E\{A_c(t)\}$ and (2.6) would

again be quadratic in time; the analogue of (2.3a) being

$$y^2(n/8)[E\{S\}\alpha(t_0^*)](1 + c_S^2)^2 \quad \text{with} \quad y = (t - t_0^*)/E\{S_0\} , \tag{2.8}$$

for $0 < y < 1$. Although $E\{A_c(t)\}$ and (2.6) do not depend on C_S , the wider range of $t_0 - t_0^*$ causes (2.8) to increase quite rapidly with C_S for fixed y , and, in particular at $y = 1$. It is 4 times as large for exponentially distributed service times $(C_S = 1)$ than for regular service times $(C_S = 0)$. Although for $C_S = 0$, we considered (2.8) to be negligible, it may not be for $C_S = 1$. The larger value of (2.8) will, however, be more effective in cancelling some of the stochastic corrections.

After queueing starts, the correct value of $E\{D(t)\}$ is determined from the solution of II(5.2). It approaches a straight line of slope $n/E\{S\}$, within a few service times if C_S is not too small. The details of how $E\{D(t)\}$ behaves near time t_0 is of minor importance. We are more concerned here with the asymptotic difference between $E\{D(t)\}$ and (2.6) for large t .

Suppose, as in (II 5.2), we could evaluate $E\{D(t)\}$ until some time τ_1 when it was virtually certain that $N(\tau_1) > 0$. From (II 5.2) one can show that if $\varepsilon(t)$ is the deviation of $E\{D(t)\}$ from any straight line of slope $n/E\{S\}$, i.e.

$$\varepsilon(t) = E\{D(t)\} - tn/E\{S\} - a \tag{2.9}$$

for any constant a , then the asymptotic limit of $\varepsilon(t)$ is given by

$$\varepsilon(\infty) = \int_0^\infty d\tau \, \frac{\overline{G(\tau)}}{E\{S\}} \, \varepsilon(\tau_1 - \tau) = \int_0^\infty dG_0(\tau)\varepsilon(\tau_1 - \tau) , \tag{2.10}$$

where $G_0(\tau)$ is the distribution given in (II 2.2). Thus the final asymptotic deviation from the line is an average, weighting $dG_0(\tau)$ of deviations at times prior to τ_1 .

This result is, of course, derived from (II 5.2) under the assumption that II(5.2) applies for all $t > \tau_1$. It actually applies only for a finite time until the queue vanishes again, but, if the queue persists for many service times, we expect $\varepsilon(t)$ to approach the limit $\varepsilon(\infty)$ well within the time of queueing.

If, again, we neglect stochastic queueing effects over the transition, then we may choose $\tau_1 = t_0$ in (2.10) and take $\varepsilon(t)$ equal to (2.8) for $t < t_0$, thus

$$\varepsilon(\infty) \;=\; \int_0^\infty dG_0(\tau)(E\{S_0\} - \tau)^2 n\alpha(t_0^*)/2E\{S\}$$

$$= \left(\frac{n}{2}\right)[\alpha(t_0^*)E\{S\}] \frac{Var\{S_0\}}{E^2\{S\}} \; . \tag{2.11}$$

This is the amount by which, in the absence of the stochastic queueing, (2.6) under-estimates $E\{D(t)\}$ after several service times.

The effects of the random service appear in the factor $Var\{S_0\}/E^2\{S\}$. For (nearly) regular service, S_0 would be uniformly distributed over $(0 , s)$ and $Var\{S_0\}/E^2\{S\}$ would have a value of $1/12$, in agreement with the average value (2.3b). For exactly regular service $\varepsilon(t)$ actually does not approach a limit for $t \to \infty$, but for a narrow distribution one will obtain approximately this value of $1/12$ (with a correction term proportional to C_S) . The value of $\varepsilon(\infty)$, however, is quite sensitive to the service distribution. For exponentially distributed S , S_0 has the same distribution as S , and $Var\{S_0\}/E^2\{S\} = 1$, <u>twelve</u> times as large as for nearly regular service. The value of $\varepsilon(\infty)$ is, in fact, the same as $\varepsilon(t_0)$ at the time the queueing was assumed to start.

Whereas for regular service we considered $\varepsilon(\infty)$ to be of negligible importance, it may be important for systems comparable with exponentially distributed service. The exponential service actually has special properties not immediately apparent from equations such as (II 5.2). It is no accident that the value of $\varepsilon(\infty)$ was equal to $\varepsilon(t_0)$. Because of the "memoryless" property of the exponential distribution, the rate of service is exactly $n/E\{S\}$ for all $t > \tau_1$ in (II 5.2), not just asymptotically after many service times. Thus, if we neglect stochastic queueing, $\varepsilon(\infty)$ is the same as at $\tau_1 = t_0$, and the curve for $E\{D(t)\}$ takes off from the curve $E\{A_c(t)\}$ at time t_0 with a slope of $n/E\{S\}$, rather than at time t_0^* .

It is interesting that, for fixed $E\{A_c(t)\}$ and $E\{S\}$, the value of $\varepsilon(\infty)$ is an increasing function of the variation in S . This means that, in the absence of the stochastic correction (II 3.10c), the expected queue is a <u>decreasing</u> function of C_S ; the larger $\varepsilon(\infty)$ the higher is the curve $E\{D(t)\}$. Furthermore, the effect may be quite significant in certain situations.

The above results were restricted by the neglect of stochastic queueing effects over the transition. According to (II 2.18), there is an expected excess of servers at time t_0^* equal to

$$- E\{N(t_0^*)\} \simeq n - E\{S\}\lambda(t - E\{S_0\})$$
$$\simeq nE\{S_0\}\alpha(t_0^*)$$

provided that this is larger than approximately $\sigma_n \simeq (I*n)^{1/2}$ of (II 2.21). This condition

$$[E\{S\}\alpha(t_0^*)] \gtrsim 2(I*/n)^{1/2}[1 + c_S^2]^{-1} \tag{2.13}$$

also guarantees that stochastic queueing will exist for at most a time comparable with $E\{S\}$ so that queueing effects do not have much time to accumulate.

Equation (2.13) is the generalization of (2.5). If $I*$ and I are nearly equal, the factor $[1 + c_S^2]^{-1}$ in (2.13) may extend somewhat the range of $\alpha(t_0^*)$ for which stochastic effects are negligible at time t_0^* . If, however, I is small compared with 1 , $I*$ may be considerably larger than I (unless c_S^2 is also small).

If (2.13) is true (preferably with a factor of 2 or 3 difference between the two sides), the stochastic correction can be estimated from (II 3.10c). It predicts a maximum correction to $E\{D(t)\}$ at time t_0 of about $0.4\sigma_N$, in the opposite direction to the $\varepsilon(t)$ correction described above. To estimate the effect of this on $\varepsilon(\infty)$, one should also substitute the correction term of (II 3.10c) into (2.10). If the transition period is short compared with $E\{S\}$, one can choose τ_1 sufficiently close to t_0 that $\overline{G}(\tau)$ in (2.10) is close to 1 over the transition. Thus the correction to $\varepsilon(\infty)$ is approximately

$$- \int_0^\infty d\tau(\sigma_N/E\{S\})H\{- |E\{N(\tau_1 - \tau)|/\sigma_N\}\} . \tag{2.14}$$

In the vicinity of the transition

$$E\{N(\tau_1 - \tau)\} \simeq n(\tau_1 - \tau)\alpha(t_0)$$

varies linearly with time and so (2.14) becomes

$$- \frac{2\sigma_N^2}{n[\alpha(t_0)E\{S\}]} \int_0^\infty dz H(-z) \;=\; - \frac{I^*}{2[\alpha(t_0)E\{S\}]} \cdot$$

The corrected form of $\varepsilon(\infty)$ is therefore

$$\varepsilon(\infty) \;\simeq\; \left(\frac{n}{2}\right)[\alpha(t_0^*)E\{S\}] \frac{\mathrm{Var}\{S_0\}}{E^2\{S\}} - \frac{I^*}{2[\alpha(t_0)E\{S\}]} \cdot \tag{2.15}$$

This approximation is certainly valid if $G(\tau_1 - \tau_0) \ll 1$ as in (II 5.1), and it should be a fair estimate as long as (2.13) is true. The formula, because of the way it was derived, contains $\alpha(t_0^*)$ is one term and $\alpha(t_0)$ in the other, but $\alpha(t)$ is assumed to be nearly constant between t_0^* and t_0. We can also use $\alpha(t_0^*)$ in the second term of (2.15).

From (2.15) we see that $\varepsilon(\infty) \geq 0$, i.e. the deterministic correction (2.11) exceeds to stochastic correction if

$$[\alpha(t_0^*)E\{S\}] \;\gtrsim\; \left[\frac{I^*}{n} \frac{E^2\{S\}}{\mathrm{Var}\{S_0\}}\right]^{1/2} \cdot \tag{2.16}$$

In the special case of regular service, $S = s$, $E^2\{S\}/\mathrm{Var}\{S_0\} = 12$, and the right hand side of (2.16) gives $(12)^{1/2}(I/n)^{1/2} \simeq (3.5)(I/n)^{1/2}$. We had previously esti- mated $(3.2)(I/n)^{1/2}$ from the values of the correction terms at time t_0 . For ex- ponential service, the right hand side of (2.16) is $(I^*/n)^{1/2} = [(I + 1)/2n]^{1/2}$.

The condition (2.16) is quite similar to the condition (2.13). We do not know which of these better defines the range of validity of (2.15), but in any case where there is some doubt about the accuracy of (2.15), the value of $\varepsilon(\infty)$ is likely to be so small as to be of no practical importance anyway.

We will not pursue the details of this further here, because for values of n in the range of 10 to 20, (2.13) is not likely to be satisfied in most of the in- tended applications. If $E\{S\}$ is small enough to justify the use of (1.1), it is likely also that stochastic effects accumulate over the transition and overpower the deterministic correction in (2.11). In any case, it is convenient to adopt the simple stright line (2.6) as a first approximation and then worry about whether this approximation underestimates or overestimates the correct $E\{D(t)\}$.

3. <u>Small Queues</u>. Having by now systematically exhausted the analysis of most situ- ations in which stochastic effects (if any) persist for at most about an average

service time, we must finally face the problem in which this is not true, i.e. $E\{S\}$ is not just small compared with the duration of the rush hour, but is comparable with or small compared with the time during which it is uncertain as to whether $N(t)$ is positive or negative.

Again there are two distinct rush hour situations to consider. First, $\rho(t)$ increases to a maximum value close to 1 so as to cause $N(t)$ to pass from a state in which it is certain that $N(t) < 0$ for $t < \tau_0$ to a state when it is again certain that $N(t) < 0$ for $t > \tau_1 > \tau_0$ but for $\tau_0 < t < \tau_1$, the distribution of $N(t)$ overlaps 0. In the second case, $\rho(t)$ exceeds 1 so that the state goes to $N(t) > 0$ for $t > \tau_1$. In this case, $N(t)$ must eventually decrease again and pass through a second transition from $N(t) > 0$ to $N(t) < 0$.

In previous sections we have been working mostly with the curve $E\{A_c(t)\}$ and other curves derived from it, and we are further assuming here that $E\{A_c(t)\}$ is smooth (approximately linear or quadratic) over times of order $E\{S\}$ or over any transitions. Up to now, we have found that, although any realization of $A_c(t)$ will have stochastic irregularities, these irregularities had little influence on $E\{N(t)\}$. Even when we drew graphs of realizations $A_c(t)$ as in Fig. I.4, we drew them as if $A_c(t)$ were smooth.

This has caused no misconceptions yet because of the following: (1) Prior to the time when queueing starts, we were always taking expectations of sums or differences and, because of the linearity of the expectation, any stochastic effects average out. (2) The same was true after queueing becomes certain, at least for the evaluation of changes in queue lengths, departures, etc., from one time to another. (3) During the transition period, realizations of $N(t)$ were determined by the total number of arrivals during a previous service time. Because $N(t)$ is itself a running average of differences in $A_c(t)$, a single realization of $N(t)$ will remain nearly constant over times short compared with $E\{S\}$, although the value of $N(t)$ will vary from one realization to another. Thus at the only time when the time variation of $N(t)$ could have been important, $N(t)$ was nearly constant over the time periods in question (which were short compared with $E\{S\}$.

The properties (1) and (2) continue to be valid as we consider small values of

$E\{S\}$, but not the third property. The time dependence of individual realizations

of $A_c(t)$ and $N(t)$ will be important.

Before we get into the main problem of estimating the accumulation of stochastic

effects over the transition, we consider here some situations with $1 - \rho(t) \gtrsim \sigma_N/n$

in which some formulas derived in the previous sections will apply, but for new

reasons.

Suppose that $1 - \rho(t)$ maintains a value of about $\sigma_N/n = (I*/n)^{1/2}$ for at

least several mean service times. The traffic intensity $\rho(t)$ may either increase

to a maximum of about $1 - \sigma_N/n$, staying near this maximum for several service times

or $\rho(t)$ may increase very slowly (eventually it may exceed one), but we are con-

cerned only with the behavior of $N(t)$ before $\rho(t)$ gets much beyond $1 - \sigma_N/n$.

During this time, a single realization of $A_c(t)$ may have a section with a shape

like that shown in Fig. II.4, even though, over this range of time, $E\{A_c(t)\}$ may

be nearly linear with a slope λ . A queue forms between about points 2 and 3 of

Fig. II.4 (for S = s) which in turn retards the curve $A_s(t)$ of available servers

at 2" and 3". But at points 2" and 3", there was an excess of servers and the dis-

tortion of the curve $A_s(t)$ has no lasting effects.

In Section II.3, if $\rho(t)$ were to increase to a maximum, a queue would last

for a time less than $E\{S\}$ because $\rho(t)$ was larger than $1 - \sigma_N/n$ only for a

time less than $E\{S\}$. Here a queue, if it exists, will last for a time less than

$E\{S\}$ because the fluctuation in $A_c(t)$ which caused the queue will decay in a time

less than $E\{S\}$.

To analyse the consequences of this, consider again the special case S = s .

If, at time t , $1 - \rho(t - s) \gtrsim \sigma_N/n$, then the effects of the last term of (I 2.11)

will be small because $N(t - s)$ is usually negative. The first approximation to the

distribution of $N(t)$ would be a normal distribution with a mean

$$E\{N(t)\} \simeq s\lambda(t - s/2) - n = n[\rho(t - s/2) - 1] , \qquad (3.1)$$

as in (II 2.6b) , and a variance of $\sigma_N^2 \simeq In$. The distribution of the customer

queue, $N_c(t) = \max [0 , N(t)]$ will then be approximately a truncated normal dis-

tribution with a mean value given by (II 3.7)

$$E\{N_c(t)\} \;\simeq\; \sigma_N H([\rho(t - s/2) - 1]n/\sigma_N) \; . \tag{3.2}$$

One can test the validity of this approximation, or evaluate a second approximation, by substituting a normal distribution for $N(t - s)$ back into (I 2.11) and reevaluating the distribution of $N_c(t)$. If the queue length at time $t - s$, the second term of (I 2.11), is small compared with the standard deviation of the first term, i.e.

$$E\{N_c(t - s)\} \;<<\; \sigma_N \;\simeq\; (In)^{1/2} \tag{3.3}$$

then the second approximation to the distribution of $N(t)$ will also be a normal distribution. This follows from the fact that the new arrivals (first term of (I 2.11)) and the queue at time $t - s$ are statistically independent. If one adds a small random variable, $N_c(t - s)$, to a larger normally distributed random variable, one obtains another approximately normal random variable (independent of the distribution of the queue), with a mean and variance equal to the sum of the means and variances, respectively, of the two random variables. Thus the second approximation to the distribution of $N(t)$ is a normal distribution with mean

$$
\begin{aligned}
E\{N(t)\} \;&\simeq\; [s\lambda(t - s/2) - n] + E\{N_c(t - s)\} \\
&\simeq\; n[\rho(t - s/2) - 1] + Var\{N(t - s)\}]^{1/2} H\left(\frac{E\{N(t - s)\}}{[Var\{N(t - s)\}]^{1/2}} \right)
\end{aligned}
\tag{3.4}
$$

and

$$
\begin{aligned}
\sigma_N^2 \;&=\; Var\{N(t)\} \simeq Is\lambda(t - s/2) + Var\{N_c(t - s)\} \\
&\simeq\; In\rho(t - s/2) + Var\{N(t - s)\} H*\left(\frac{E\{N(t - s)\}}{[Var\{N(t - s)\}]^{1/2}} \right)
\end{aligned}
\tag{3.5}
$$

as in (II 3.14) and (II 3.15). The mean queue length at time t is then

$$E\{N_c(t)\} \;\simeq\; \sigma_N H(E\{N(t)\}/\sigma_N) \; . \tag{3.6}$$

Formally, these equations determine the means and variances at time t in terms of those at time $t - s$. Similar equations were used in Section II.3 to describe the evolution of $N(t)$ through and beyond the transition. We have already assumed in (3.4) to (3.6) that $A_c(t)$ is slowly varying and have approximated $A_c(t) - A_c(t - s)$ by $\lambda(t - s/2)$, but the main point of departure from the approximations of Section II.3 concerns the variation of the second terms of (3.4) and (3.5) over times of order s . In Section II.3, we were concerned with situations in which one passed through the transition in a time less than s . If $E\{N(t)\}/\sigma_N$ in (3.6) were comparable with 1, time t was in the transition but $t - s$ was well before the transition. Therefore H and H^* in (3.4) and (.5) were negligible. Here we are concerned with the opposite situation; $\rho(t)$ varies so slowly that $E\{N(t)\}/\sigma_N$ remains close to about -1, say, for several service times.

Our present postulates imply that the second terms of (3.4) and (3.5) are relatively small and slowly varying with time. It suffices therefore to use (3.1) and $\mathrm{Var}\{N(t - s)\} \simeq In$ in estimating these terms. Thus

$$E\{N(t)\} \;\simeq\; (In)^{1/2} \left\{ \frac{[\rho(t - s/2) - 1]}{(I/n)^{1/2}} + H \left(\frac{[\rho(t - 3s/2) - 1]}{(I/n)^{1/2}} \right) \right\} \qquad (3.4a)$$

and

$$\mathrm{Var}\{N(t)\} \;\simeq\; In \left\{ 1 + [\rho(t - s/2) - 1] + H^* \left(\frac{[\rho(t - 3s/2) - 1]}{(I/n)^{1/2}} \right) \right\} \qquad (3.5a)$$

If $1 - \rho(t - 3s/2) \gtrsim (I/n)^{1/2} \simeq \sigma_N/n$, both H and H^* in (3.4a) and (3.5a) will be small compared with 1. The second approximation to (3.6) will therefore be consistent with the first approximation (3.2). We will be examining the numerical accuracy of these formulas later in Chapter IV. Our main objective here was to establish the fact that, if $\rho(t)$ stays less than about $1 - \sigma_N/n$, the length of time it stays there is not critical to the validity of (3.2). Indeed if $\rho(t)$ were to remain constant at any value $\rho < 1$, the distribution of $N(t)$ should eventually approach a stationary equilibrium distribution. This is the aspect of queueing theory with which most of the literature on queueing theory deals. It is essentially the opposite situation to that discussed in previous sections. In the present case with $\rho(t) \leq 1 - \sigma_N/n$, $N(t)$ will reach an equilibrium distribution within a time s if the queue at time t is small, regardless of the previous behavior.

Furthermore, the distribution of $N(t)$ will remain approximately normal whether $\rho(t)$ varies rapidly as in previous sections or remains constant.

It will be useful also to have some estimates of the number of idle servers $E\{N_s(t)\}$. One method of obtaining this is to evaluate it directly from the distribution of $N(t)$. Since $N(t)$ is approximately normal (even in the second approximations), (II 3.10b) gives

$$\begin{aligned} E\{N_s(t)\} &\simeq - E\{N(t)\} + \sigma_N H(E\{N(t)\}/\sigma_N) \\ &\simeq - E\{N(t)\} + E\{N_c(t)\} \ . \end{aligned} \tag{3.7}$$

If we substitute the first approximation (3.1) to $E\{N(t)\}$ into (3.7), we obtain

$$E\{N_s(t)\} \simeq n[1 - \rho(t - s/2] + E\{N_c(t)\} \ , \tag{3.7a}$$

but if we substitute the second approximation (3.4), we obtain

$$E\{N_s(t)\} \simeq n[1 - \rho(t - s/2)] - E\{N_c(t - s)\} + E\{N_c(t)\} \ . \tag{3.7b}$$

The first terms of (3.7a) or (3.7b) represent the number of servers that would be idle if there were no queueing; $n - E\{N_s(t)\}$ is the number of busy servers needed to serve the customers arriving at rate $\lambda(t)$. If $E\{N_c(t - s)\}$ were zero, (3.7a) and (3.7b) would agree; but if $E\{N_c(t)\}$ is slowly varying, the correction term in (3.7a) is not meaningful; the second and third terms of (3.7b) will nearly cancel. In particular, if one were dealing with an equilibrium situation, these terms would exactly cancel. The expected number of idle servers would be $n[1 - \rho]$; the expected number of busy servers would be $n\rho$. This is not just an approximate result, it is exactly true. It follows from the equilibrium formula $"L = \lambda W"$, $"L"$ being the number of customers in service, and $"W"$ the service time s .

As a preliminary to the generalization of these results to the case of random service time, the point we wish to make here is that the first approximation to the distribution of $N(t)$ gives a correct estimate of $E\{N_c(t)\}$, but does not generally give a proper first correction to $E\{N_s(t)\}$. There is, however, a second method of deriving (3.7b).

From (I 2.9), the curves $E\{A_s(t)\}$ and $E\{D(t)\}$ are related through

$$E\{A_s(t)\} = E\{D(t - s)\} + n . \tag{3.8}$$

Also

$$\begin{aligned}
E\{N_s(t)\} &= E\{A_s(t)\} - E\{D(t)\} \\
&= E\{D(t - s)\} - E\{D(t)\} + n .
\end{aligned} \tag{3.9}$$

But since

$$E\{D(t)\} = E\{A_c(t)\} - E\{N_c(t)\} , \tag{3.10}$$

it follows that

$$\begin{aligned}
E\{N_s(t)\} &= [n + E\{A_c(t - s)\} - E\{A_c(t)\}] \\
&\quad - E\{N_c(t - s)\} + E\{N_c(t)\} .
\end{aligned} \tag{3.11}$$

If $E\{A_c(t)\}$ varies nearly linearly with t , the first term of (3.11) is the same as the first term of (3.7a) . The equation (3.11), however, is exact. Thus, by using (3.11) along with any first approximations to $E\{N_c(t)\}$ we can evaluate the correct first correction to $E\{N_s(t)\}$.

Equation (3.11) and its generalizations to random S will have some important applications later. The last two terms of (3.11) represent the rate of growth of the expected queue. The queue cannot grow unless $E\{N_s(t)\}$ exceeds what one would estimate from the deterministic approximation. This will eventually put some rather severe restrictions on how large the queue can become during the transition.

The above methods of approximation are particularly well suited to the case $S = s$, but they can be generalized for random S at some sacrifice in accuracy.

If a queue forms at some time due to stochastic fluctuations in the arrivals or the service, the effects will not be felt until about a service time later, at which time the queue is likely to have vanished, provided $\rho(t)$ is sufficiently small. On the basis of this, we can argue as in Sections II 4 and 5, that the distribution of $N(t)$ will be approximately the same as it would be without queueing, if $E\{N_c(t)\}$ is sufficiently small. It will be a normal distribution with a mean given by (II 2.18a)

$$E\{N(t)\} \simeq E\{S\}\lambda(t - E\{S_0\}) - n = n[\rho(t - E\{S_0\}) - 1] \qquad (3.12)$$

and variance $\sigma_N^2 \simeq I^*n$. Thus

$$E\{N_c(t)\} \simeq \sigma_N H([\rho(t - E\{S_0\}) - 1]n/\sigma_N). \qquad (3.13)$$

These should be approximately correct provided that

$$E\{N_c(t - \tau)\} << \sigma_N \qquad (3.14)$$

for all positive values of τ comparable with $E\{S\}$. These are the obvious generalizations of (3.1), (3.2), and (3.3).

We will also be testing the accuracy of this later. We will be able to show that this is a satisfactory first approximation, but that the deviations from the normal distribution will not generally yield another normal distribution in a second approximation, as for $S = s$, i.e., the deviations generally distort the shape of the distribution.

Although we cannot generalize the method used in (3.7) which was based upon the second approximations to the distribution of $N(t)$, we can generalize (3.11). From (II 2.11)

$$E\{N_s(t)\} = n - \int_0^\infty dG(\tau)[E\{D(t)\} - E\{D(t - \tau)\}] ,$$

and

$$E\{D(t)\} = E\{A_c(t)\} - E\{N_c(t)\} ,$$

we obtain

$$E\{N_s(t)\} = n - \int_0^\infty dG(\tau)[E\{A_c(t)\} - E\{A_c(t - \tau)\}]$$

$$- \int_0^\infty dG(\tau)[E\{N_c(t)\} - E\{N_c(t - \tau)\}] , \qquad (3.15)$$

which is the generalization of (3.11). If $E\{A_c(t)\}$ is slowly varying, the first integral gives the quantity (3.12). If the mean queue is also slowly varying, the same methods of approximation can be used on the second integral to give

$$E\{N_s(t)\} \simeq n[1 - \rho(t - E\{S_0\})] + E\{S\}dE\{N_c(t - E\{S_0\})\}/dt . \qquad (3.16)$$

This again relates $E\{N_s(t)\}$ to the rate of growth of the queue. If $\rho(t) = \rho$ is a constant, this yields $E\{N_s(t)\} = n[1 - \rho]$ as before.

From an approximate evaluation of $E\{N_c(t)\}$, one can now estimate $E\{N_s(t)\}$ from (3.16), and from these obtain corresponding estimates of $E\{D(t)\}$ and $E\{A_s(t)\}$, relative to the given $E\{A_c(t)\}$.

Equation (3.16) also applies for a single channel server $n = 1$ (with a slowly varying expected queue). In this case $N_s(t)$ can only have the values 0 or 1 , and $E\{N_s(t)\}$ represents the probability that the server is idle at time t , or the probability of zero customers in the system at time t . For equilibrium, this has the value $1 - \rho$, a well-known result. For time-dependent queueing, however, (3.6) illustrates the important point that queueing occurs not just because of stochastic excesses of customer arrivals (because these would be balanced by stochastic deficiencies). It is really due to the fact that a deficiency of customers causes a loss in service if the queue vanishes.

4. _Transition Behavior_. To understand the qualitative behavior of the queue during the transition, one should, as in the previous section, investigate the behavior of realizations of $A_c(t)$, $N(t)$, etc.

Suppose that at some time τ_0 we specify a value of $N(\tau_0)$. Between time τ_0 and some time $t > \tau_0$, $N(t)$ will increase by the number of new customer arrivals and decrease by the number of service completions.

$$N(t) - N(\tau_0) = [A_c(t) - A_c(\tau_0)] - [A_s(t) - A_s(\tau_0)] . \qquad (4.1)$$

Over some time $t - \tau_0 \gtrsim E\{S\}$, the two terms in brackets of (4.1) are generally positively correlated if $N(\tau) < 0$ during the time interval (τ_0 , t) , because an excess of customer arrivals will generate an excess of service completions (at later times). Although both terms of (4.1) have variances that increase more or less proportional to $(t - \tau_0)$, the variance of their difference will not necessarily increase with $t - \tau_0$ if $N(\tau)$ stays negative. If, however, $N(\tau) > 0$ during the time interval (τ_0 , t), whether due to past stochastic effects or because $E\{N(\tau_0)\} > 0$, the arrivals and departures will be nearly independent; their variances will add and increase approximately proportional to $(t - \tau_0)$.

These properties have already been discussed in Section II.4 in connection with the average behavior when $E\{N(t)\} \lesssim - \sigma_N$ or $E\{N(t)\} \gtrsim +\sigma_N$. The main point here is that these properties are still true starting from some given value $N(\tau_0)$ which may have been generated by some realization of past behavior. In particular, one might have an $N(\tau_0) > 0$ even though $E\{N(\tau_0)\} < 0$, or vice versa. Furthermore, if $\rho(\tau)$ is less than 1 during a time (τ_0 , t) and $N(\tau_0) > 0$ so that the queue is "expected" to decrease in some deterministic sense, $N(t)$ could be larger than $N(\tau_0)$. If fluctuations in $N(\tau)$ in the negative direction are damped when $N(\tau)$ goes negative, but positive fluctuations can be large, the net result could even cause $E\{N(t)\} > N(\tau_0)$. This is, of course, one way of explaining why, in the theory of stochastic queueing, with $\rho(t) = \rho < 1$, one obtains a stationary distribution of positive queue lengths.

One can explain at least the order of magnitude of most stochastic effects by the following very crude quantitative interpretation of the above behavior. Suppose that τ_0 is some time during the transition period when the distribution of $N(\tau_0)$ overlaps 0, and that this situation will continue during a time interval (τ_0 , t) of at least several mean service times. We also assume that prior to time τ_0, $\rho(t)$ had been (slowly) increasing; we had not yet passed the peak of the rush hour.

These conditions also imply that

$$\left|1 - \rho(\tau)\right| \le \sigma_N/n \simeq (I^*/n)^{1/2} \quad \text{for} \quad \tau_0 < \tau < t , \tag{4.2}$$

because, on the one hand, we would not have reached the transition period yet unless $1 - \rho(\tau_0)$ were less than about σ_N/n (as described in the last section), and, on the other hand, if $\rho(\tau) - 1 > \sigma_N/n$, the expected queue length would increase by about $[\rho(\tau) - 1]n > \sigma_N$ within a time $E\{S\}$ and pass through the transition within a time of order $E\{S\}$ or less. We will, in fact, be particularly interested in the behavior when the condition $\left|1 - \rho(\tau)\right| \ll \sigma_N/n$ persists for many service times,

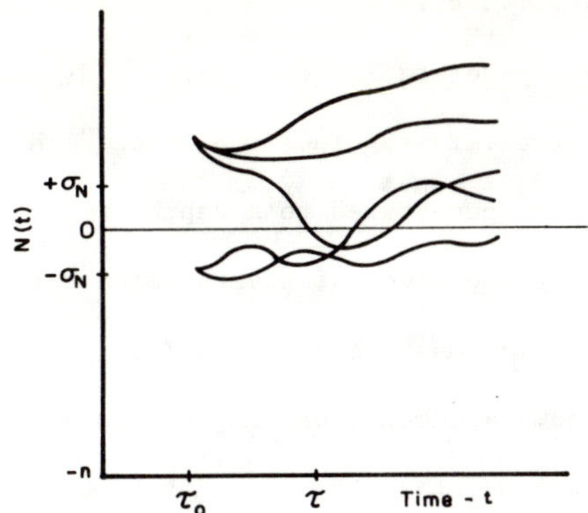

Fig. III.2 - Some typical realizations
of $N(t)$ starting from $N(\tau_0) < 0$ or
$N(\tau_0) > 0$.

because this is the situation in which the stochastic queueing effects become relatively large.

Consider now some typical realizations of $N(\tau)$ as in Fig. III.2. If
$N(\tau_0) < 0$ (or $N(\tau)$ becomes negative at some time $\tau > \tau_0$) so that there is an
excess of servers, then the system behaves temporarily as if it had an arbitrarily
large number of excess servers, and the evolution of $N(\tau)$ is essentially as described in the last section. Even if $N(\tau)$ should soon become positive, the effects
of the queue would not be felt until the queueing delayed the availability of servers at about a service time later. Within a time of order $E\{S\}$ after τ_0 , most
customers in the service will have left and been replaced by a new batch of approximately n arrivals. The conditional distribution of $N(\tau)$, given $N(\tau_0) < 0$,
would be approximately normal with a mean of about $E\{S\}\lambda(\tau_0) - n$ at time
$\tau \sim \tau_0 + E\{S_0\}$, and a variance of about σ_N^2 . This is essentially independent of the
value of $N(\tau_0)$; the system remembers its past behavior only for a time of order
$E\{S\}$, provided there was no queue at the earlier times.

If, for some particular realization of $N(\tau)$, $N(\tau)$ should remain negative for
some time (regardless of what $E\{N(t)\}$ may be doing), then it will continue to re-
generate new approximately normal conditional distributions of $N(\tau)$. It is not
possible, therefore, for the distribution of $N(\tau)$ to drift lower than that of a
normal distribution of mean approximately $E\{S\}\lambda(\tau) - n \simeq [\rho(\tau) - 1]n \gtrsim - \sigma_N$ and a
variance of about σ_N^2 . If $N(\tau)$ should become slightly positive, the queueing

will retard the availability of servers and cause an increase in $N(\tau)$ at later times. One certainly cannot generate larger negative $N(\tau)$ from past positive values than from past negative values. Thus the "state" $N(\tau) = 0$ acts like a "soft reflecting barrier;" it can be penetrated to a depth of about $-\sigma_N$, and $N(\tau)$ may even stay there for a time of the order of several mean service times even if $\rho(\tau) > 1$, but eventually $N(\tau)$ must become positive.

On the basis of the above argument, one might even go a little further and speculate that if (4.2) is true, then the shape of that part of the distribution of $N(\tau)$ with $N(\tau) < 0$, i.e., the conditional distribution of $N(\tau)$ given $N(\tau) < 0$, is quite similar to the corresponding (approximately normal) distribution of $N(\tau)$, given $N(\tau) < 0$, that would exist in the absence of queueing. The argument is that most of the distribution for $N(\tau) < 0$ is likely to have evolved from realizations for which $N(\tau)$ has been negative or close to 0 for at least one or two mean service times, during which time the service has served nearly all its customers and acquired a new approximately normal distribution for $N(\tau)$. Of course, we have made no statement here about $P\{N(\tau) < 0\}$, only on the conditional distribution, given $N(\tau) < 0$; we have guessed at the approximate shape of this distribution but not the amplitude.

If, on the other hand, $N(\tau_0) > 0$ or $N(\tau)$ becomes positive, either because $\rho(\tau) > 1$ or because of positive fluctuations at previous times, there is a significant probability that it will stay positive for several mean service times. If so, the distribution of $N(\tau)$ will spread over a range comparable with the standard deviation of (4.1), i.e., according to (II 4.2) over a range of about

$$[(1 + c_S^2)\lambda(\tau_0)(\tau - \tau_0)]^{1/2} \simeq [I^+(\tau - \tau_0)n/E\{S\}]^{1/2} , \tag{4.3}$$

provided $\lambda(\tau)$ remains nearly constant at approximately $n/E\{S\}$ over time τ_0 to τ . If, however, this distribution spreads to negative values of $N(\tau)$, it will be reflected off the soft barrier. If it stays above the barrier, it will also be centered around a time varying mean of approximately

$$N(\tau_0) + [n/E\{S\}] \int_{\tau_0}^{\tau} [\rho(x) - 1]dx . \tag{4.4}$$

In any case, if $N(\tau_0) > 0$, $E\{N(\tau)\}$ will try to decrease at time τ_0 at an expected rate $[1 - \rho(\tau_0)]n/E\{S\}$, (or increase at a rate $[\rho(\tau_0) - 1]n/E\{S\})$.

Various types of queueing behavior during the transition arise from the interplay among the effects of the soft barrier, the spread of the distribution when $N(\tau) > 0$ in (4.3), and the drift of the mean (4.4).

We have already seen in Section 3 that as long as $\rho(t) \lesssim 1 - \sigma_N/n$, the distribution of $N(t)$ remains close to the "instantaneous equilibrium distribution," i.e. the equilibrium distribution associated with the traffic intensity ρ evaluated at time t (or, more correctly, at about time $t - E\{S_0\})$. In the above interplay, the behavior within the barrier dominates. Occasionally a queue forms, but the drift (4.4) annihilates it within a time less than a mean service time. The effects of (4.3) have little influence.

If $\rho(t)$ exceeds $1 - \sigma_N/n$, but $|1 - \rho(t)| < \sigma_N/n$ for a time less than about $E\{S\}$ (either passing through or reaching a maximum), the methods of Section 2 apply. If, however, $\rho(t)$ exceeds $1 - \sigma_N/n$, increasing at a slow rate, the distribution of $N(t)$ will still try to follow that equilibrium distribution associated with a parameter $\rho(t)$.

These equilibrium distributions (which we will discuss in Chapter IV) will be maintained by a competition between the combined effects of the fluctuation (4.3) and the reflecting barrier against the negative drift of (4.4). A new characteristic time, however, now enters the problem, the "relaxation time" of the queue distribution. In essence, this is the time $(\tau - \tau_0)$ such that the drift $[1 - \rho(\tau_0)](\tau - \tau_0)n/E\{S\}$, (4.4), can overpower the fluctuations, i.e.

$$[1 - \rho(\tau_0)](\tau - \tau_0)n/E\{S\} \sim [I^{\dagger}(\tau - \tau_0)n/E\{S\}]^{1/2}$$

or

$$(\tau - \tau_0) \sim I^{\dagger}[E\{S\}/n][1 - \rho(\tau_0)]^{-2} . \tag{4.5}$$

over which time the typical size of the fluctuations are of order (4.3), i.e. of order

$$I^{\dagger}/[1 - \rho(\tau_0)] . \tag{4.6}$$

If $1 - \rho(\tau_0)$ is comparable with $\sigma_N/n = (I^*/n)^{1/2}$, and $I^\dagger$ is comparable with I^* , the time (4.5) will be of order $E\{S\}$, and (4.6) of order $n^{1/2}$.

As $1 - \rho(\tau_0)$ decreases, the typical queue length (4.6) increases, but so does the time (4.5); it takes a long time to generate large stochastic queues. For $1 - \rho(\tau_0) << (I^\dagger/n)^{1/2}$, the time (4.5) also becomes large compared with the relaxation time of about $E\{S\}$ for the distribution of negative $N(\tau_0)$. Although the distribution of negative $N(\tau_0)$ can maintain its equilibrium shape (more or less), it might not be possible for the queue to grow fast enough to keep up with the equilibrium distribution for $N(\tau_0) > 0$. If $\rho(t)$ increases to a maximum less than 1, at a sufficiently slow rate, the distribution of $N(t)$ might be able to stay close to the instantaneous equilibrium distribution at all times, but if $\rho(t)$ gets too close to 1 or passes through 1, it will not.

If $\rho(t)$ passes through 1 at time t_0 at a rate $\alpha(t_0)$ as in Section 2, then the competition near time $\tau_0 \sim t_0$ will be between the fluctuations (4.3) and a drift term from (4.4) of order

$$[n/E\{S\}] \int_{t_0}^{\tau} x\alpha(t_0)dx = \frac{1}{2} [n/E\{S\}](\tau - t_0)^2\alpha(t_0) \ .$$

These two will be of comparable size if

$$[n/E\{S\}](\tau - t_0)^2\alpha(t_0) \ = \ [I^\dagger(\tau - t_0)n/E\{S\}]^{1/2}$$

or

$$(\tau - t_0) \sim (I^\dagger/n)^{1/3}E\{S\}[\alpha(t_0)\alpha\{S\}]^{-2/3} \tag{4.7}$$

during which time the fluctuations (the queue length) will be of order

$$I^{\dagger 2/3}n^{1/3}[\alpha(t_0)E\{S\}]^{-1/3} \ . \tag{4.8}$$

The time (4.7) is to be interpreted as a measure of the time required for the system to pass through the transition, from a state prior to time t_0 when the queue was small compared with (4.8) to a state after time t_0 when the queue is almost certain to be positive. This applies, however, only if (4.7) is large compared with $E\{S\}$, i.e.

$$\alpha(t_0)E\{S\} \ll (I^\dagger/n)^{1/2} \, , \qquad\qquad (4.9)$$

and $\alpha(t_0)$ remains nearly constant over the transition.

Except possibly for some differences between $I^\dagger$ and $I*$, this condition is essentially the opposite condition to (2.13) or (2.16) in Section 2; the stochastic effects will completely dominate the deterministic correction of Section 2. The conditions (4.9) also implies that the typical queue length (4.8) is "large" compared with $(I^\dagger n)^{1/2}$. Since the latter is the typical queue length while $1 - \rho(t)$ is comparable with σ_N/n , this means that, if (4.9) is true, the system will indeed try to follow a near equilibrium distribution until the queue has grown from order $(I^\dagger n)^{1/2}$ to the order of (4.8). This may not be very long, however, because (4.8) is proportional to only the $(-1/3)$ power of $[\alpha(t_0)E\{S\}]$, a very slowly varying function. Even if (4.9) is true, (4.8) may not be very much larger than $(I^\dagger n)^{1/2}$,

The case in which $\rho(t)$ reaches a maximum near 1 is more complicated. In addition to the relaxation time $E\{S\}$ for the distribution of negative $N(\tau)$, and the characteristic time (4.6) for the maximum value of $\rho(\tau_0)$, if it is less than 1, there is still a third time parameter. If $\rho(\tau_0)$ is sufficiently close to 1 the queue behavior will be dominated by a competition between the fluctuations (4.3) and a term of (4.4) from (1.2a) of the form

$$[n/E\{S\}]\tfrac{1}{2} \int_{\tau_0}^{\tau} x^2 \beta(\tau_0)dx \;=\; [n/E\{S\}]\tfrac{1}{6}(t - \tau_0)^3 \beta(\tau_0) \; .$$

We will examine some of these cases in more detail in Chapter IV. The method we will use to apply diffusion approximations to analyse the behavior of that part of the distribution with $N(\tau) > 0$, to apply some normal distributions for $N(\tau) < 0$, and then try to piece the two together. As long as $N(t)$ stays positive for a time large compared with $E\{S\}$, the queue behavior for n servers is nearly the same as for one server with mean service time $E\{S\}/n$. The diffusion approximations for the single channel server have already been analyzed for the various types of transition behavior described above [10, 11] . We need only try to make some corrections for a soft vs. a hard barrier.

There will still be a gap between these approximations and those described in

Section 2, for situations in which the duration of the transition is just a few mean service times. Hopefully the gap will not be so large that one cannot make a crude interpolation between them (stochastic queues do not increase very rapidly as the duration of the transition increases).

5. <u>The final transition</u>. If $\rho(t)$ exceeds 1 so that the first transition leads to a state in which a queue is almost certain, there must be a second transition at the end of the rush hour back to $N(t) < 0$. Unlike the corresponding cases discussed in Chapter II, this transition may also last for a time larger than $E\{S\}$. But, whereas the duration of the first transition depended upon the fluctuations in the arrivals during the transition, the duration of the second transition is likely to depend mostly on the cumulative fluctuations in arrivals and service during the whole period of queueing, and be rather insensitive to fluctuations during the transition itself.

If there is a "second transition," there must be a time t_1 just before the start of this transition such that $N(t_1) > 0$ is almost certain. Furthermore $\rho(t_1) < 1$, for otherwise $E\{N(t_1)\}$ was increasing before t_1 and we were still going through the first transition just before t_1 . We also assume that $\rho(t)$ is decreasing (or at least non-increasing) for $t > t_1$; we are not about to start another rush hour.

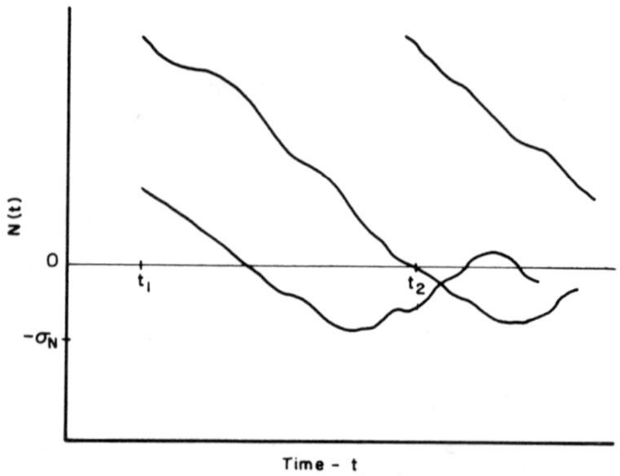

Fig. III.3 — Some typical realizations of $N(t)$ during the second transition.

Fig. III.3 shows some hypothetical realizations of $N(t)$ over the second transition. Starting at time t_1 with some value of $N(t_1) > 0$, $E\{N(t)\}$ decreases at a rate $[1 - \rho(t)]n/E\{S\}$ as in (4.4), as long as $N(t)$ stays positive. Although any realization of $N(t)$ may increase at various times, $N(t)$ will eventually become negative.

If at some time $t_2 > t_1$, $N(t_2) = 0$, then the distribution of $N(t)$ given $N(t_2) = 0$ will become approximately normal within a time of order $E\{S\}$ after time t_2 , with a mean of about $n[\rho(t_2) - 1]$ as in (3.12), and a variance $I*n$. If $\rho(t_2) \lesssim 1 - (I*/n)^{1/2}$, this distribution will be near the instantaneous equilibrium distribution. For $\rho(t)$ decreasing in time, this distribution of $N(t)|N(t_2) = 0$ will remain approximately normal with the properties described in Section 3. For all practical purposes, the queueing ceases as soon as $N(t)$ reaches zero, if, at the time, $\rho(t) \lesssim 1 - (I*/n)^{1/2}$. If, however, at time t_2 , $\rho(t) \lesssim 1 - (I*/n)^{1/2}$, then the distribution of $N(t)|N(t_2) = 0$ will try to adjust to whatever the equilibrium distribution may be.

The second transition is considered to last until a time t_f when $N(t_f) < 0$ is almost certain. This time may be determined primarily by (a) the distribution of $N(t_1)$, or (b) the time when the equilibrium queue length becomes negligible, i.e. $\rho(t_f) \simeq 1 - (I*/n)^{1/2}$, or (c) the magnitude of the queues (of order $I^{\dagger}/[1 - \rho(t)]$) for somes t , $t_1 < t < t_f$.

Case (a) will certainly apply if

$$P\{N(t_1) \lesssim I^{\dagger}/[1 - \rho(t_1)]\} \ll 1 \tag{5.1}$$

i.e. $N(t_1)$ is almost always larger than the equilibrium queue length at time t_1 . In this case, queueing ceases as soon as nearly all realizations of $N(t)$ have reached zero for the first time.

If, at the other extreme

$$P\{N(t_1) \gtrsim I^{\dagger}/[1 - \rho(t_1)]\} \ll 1 , \tag{5.2}$$

i.e., $N(t_1)$ is seldom larger than a typical equilibrium queue length at time t_1 , then the magnitude of subsequent queue lengths will depend mostly on the size of stochastic queues generated after time t_1 and be nearly independent of $N(t_1)$. If now, $\rho(t)$ changes slowly enough so that the distribution of $N(t)$ can catch up with the equilibrium distribution before $\rho(t)$ becomes less than $1 - (I^{\dagger}/n)^{1/2}$, this is when queueing will effectively cease, otherwise queueing ceases when any queue created during the transition will have been served.

These latter cases, for which (5.2) is true, are of practical importance, but, since the queue at time t_1 is less than the equilibrium queue length, the queue is expected to reach its largest value at some time after t_1 . One should, perhaps, classify this as a variation on the "first transition" rather than a special case of a "final transition."

A typical situation in which (5.2) might be true is the following. Suppose the arrival process of customers can be considered as the superposition of two parts $\rho(t) = \rho_1(t) + \rho_2(t)$. One of these, $\rho_1(t)$, is a slowly varying function which reaches a maximum value with $\rho_1(t)$ close to 1 as described in previous sections, the maximum occurring just before the time t_1 . The queue that would be generated by these customers alone, however, would not have reached an equilibrium by time t_1 Even though at time t_1 , $\rho(t_1) < 1$ and $\rho(t_1)$ is decreasing, the expected queue length is still increasing, due to fluctuations in the arrival process, in an attempt to catch up with the equilibrium distribution.

The second component of the arrival stream varies much more rapidly than $\rho_1(t)$ Although it is possibly still slowly varying on a time scale of $E\{S\}$ in the sense of (1.2a), relative to the time scale of $\rho_1(t)$ it appears like an impulse. This impulse will end just before time t_1 and contain perhaps a total of only about $(I*n)^{1/2}$ customers, enough, however, so that, when added to $\rho_1(t)$, it causes $N(t_1) > 0$ to be virtually certain at time t_1 . This creates a "second transition" in the sense that $N(t_1) > 0$ is certain, $N(t) < 0$ will become possible for $t > t_1$, and eventually $N(t) < 0$ will become certain without again going through a stage in which queueing is certain. It would obviously be more convenient, however, to consider this situation as a perturbation on the behavior with just the one component $\rho_1(t)$.

An example of a real situation in which this might happen is at a toll plaza. The component $\rho_2(t)$ is a relatively small component of the traffic stream which originates from a nearby factory. It gives a surge of traffic (of duration large compared with the service time of a few seconds) at approximately the same time every day, but of a magnitude comparable with the uncertainty in the traffic stream coming from the $\rho_1(t)$. The surge may last for a few minutes, but the toll plaza is congested for a half hour or so.

If $\rho(t)$ is smooth, not just on a time scale of $E\{S\}$, but on a time scale comparable with the time required to go through the transition (or over the whole rush hour) and if queueing becomes certain at some time when $\rho(t) > 1$, then we would expect that (5.1) is true. In going through the first transition, $\rho(t)$ was greater than 1. The first transition does not end until $N(t) > 0$ becomes certain, at which time $\rho(t)$ must still be greater than 1. The second transition will not start until $\rho(t)$ has first decreased to 1, then dropped below 1 and stayed below 1 long enough to serve enough of the queue generated before t_1 that $N(t)$ might go negative soon after time t_1 .

From the results of the diffusion approximation [10, 11], one can show that by the time the first transition is complete, i.e. $N(t) > 0$ becomes virtually certain, the distribution of $N(t)$ is already approximately normal (at least if $\rho(t)$ varies nearly linearly with t over this transition). As long as queueing continues, changes in $N(t)$ are also approximately normal, so that the queue distribution becomes even closer to a normal distribution. By the time the second transition starts, it is therefore safe to assume that the distribution of $N(t_1)$ will be normal with $E\{N(t)\}$ equal to perhaps $2[\mathrm{Var}\{N(t_1)\}]^{1/2}$, so that $N(t_1) > 0$ is virtually certain. At this time $\mathrm{Var}\{N(t_1)\}$ is comparable with $I^{\dagger}$ times the expected number of arrivals since queueing started, i.e.

$$\mathrm{Var}\{N(t_1)\} \sim I^{\dagger}[E\{A_c(t_1)\} - E\{A_c(t_0)\}] . \tag{5.3}$$

If $N(t_1)$ is approximately normal at time t_1 and satisfies (5.1), we can obtain some crude estimates of the distribution of $N(t)$ for $t > t_1$. First let $N^*(t)$ represent the process that would exist if, after time t_1 , all servers could be put back into service as soon as they were free, even if there were no customers to serve (i.e. $N(t) < 0$) . The distribution of $N^*(t_1)$ will be the same as $N(t_1)$ but $N^*(t)$ will continue to have a normal distribution for all $t > t_1$ with

$E\{N^*(t)\}$ decreasing at a rate $[1 - \rho(t)]n/E\{S\}$ and $Var\{N^*(t)\}$ increasing at a rate $I^{\dagger}\rho(t)n/E\{S\}$, even after $E\{N^*(t)\} < 0$.

Of course, $N^*(t)$ is stochastically less than $N(t)$, i.e.

$$P\{N(t) > z\} > P\{N^*(t) > z\}$$
$$\simeq 1 - \Phi([z - E\{N^*(t)\}]/Var\{N^*(t)\}]^{1/2}) \; . \tag{5.4}$$

But, if, for some $z \gg I^{\dagger}/[1 - \rho(t)]$, $N(t) > z$, then it is highly unlikely that $N(t)$ could have reached this value by virtue of having crossed 0 at some time t_2 with $t_1 < t_2 < t$. Thus for most realizations leading to $N(t) > z$, $N(t) = N^*(t)$, and the two sides of (5.4) are nearly equal.

For z less than about $I^{\dagger}/[1 - \rho(t)]$, or particularly for $z < 0$, the two sides of (4.5) may be quite different. But from the descriptions of Fig. III.3, we can make some crude corrections to (5.4). If $N^*(t) < 0$, $N(t_2)$ must have been zero at some $t_2 < t$ (when $N^*(t_2)$ vanished for the first time). Whereas $N^*(\tau)$ follows some evolution leading from $N^*(t_2) = 0$ to $N^*(t) < 0$, $N(\tau)|N(t_2) = 0$ tries first to adjust to a normal distribution that would exist in the absence of further queueing and later to the equilibrium distribution, if queueing reoccurs.

If at each time t , we were to take the probability $P\{N^*(t) < 0\}$ and redistribute it according to the equilibrium distribution at time t , we would generally tend to overestimate $N(t)$. We assume that the distributions are changing slowly enough that we need not worry about the fact that distributions at one time depend upon arrivals, service, etc., at times of order $E\{S\}$ earlier. A more important source of error should arise because $P\{N^*(t) < 0\}$ is not quite the same as the probability that $N(t)$ crossed 0 at some time $\tau < t$ and started its attempt to reach an equilibrium distribution (the latter is larger). On the other hand, it takes some time for $N(t)|N(t_2) = 0$ to reach an equilibrium distribution, during which time $N(t)|N(t_2) = 0$ is stochastically less than its equilibrium value. These two errors are in opposite directions but if $N(t_2) = 0$, $N^*(t)$ is almost certain to be negative before $N(t)$ can reach an equilibrium. The errors due to the above approximation do not accumulate, but at any particular time it would seem that the second source of error (the failure to reach an equilibrium) would be the larger error.

If we let $F_0(z, \rho)$ denote the equilibrium distribution of $N(t)$ corresponding to a traffic intensity $\rho < 1$,

$$F_0(z ; \rho) = P\{N(t) < z ; \rho(t) = \rho\} , \tag{5.5}$$

then our revised estimate of the distribution of $N(t)$ is

$$P\{N(t) > z\} \simeq \begin{cases} P\{N^*(t) > z\} + P\{N^*(t) < 0\}[1 - F_0(z ; \rho)] & \text{for } z > 0 \\ P\{N^*(t) > 0\} + P\{N^*(t) < 0\}[1 - F_0(z ; \rho)] & \text{for } z < 0 . \end{cases} \tag{5.6}$$

This is continuous at $z = 0$ but the probability density has a false discontinuity at $z = 0$ generated from interpreting $N(t) = 0$ as a critical boundary.

We would expect the right hand side of (5.6) to be too large if the equilibrium distribution $F_0(z ; \rho)$ has such a long tail that $N(t)$ could not have reached the equilibrium distribution. If so, another approximation would be to replace the $F_0(z ; \rho)$ in (5.6) by the normal distribution of (3.12) with no queueing. This would certainly make the right hand side of (5.6) too small.

The distribution (5.6) at least describes the correct qualitative behavior. As one passes through the transition, $E\{N^*(t)\}$ decreases and eventually becomes negative. The first term of (5.6) for $z > 0$ decreases and eventually becomes negligible when $E\{N^*(t)\} \lesssim - [\text{Var } N^*(t)]^{1/2}$. At all times the distribution of $N(t)$ is cut off at negative values of z by the equilibrium distribution of $N(t)$ which will not allow $N(t)$ to penetrate less than about $-(I^*n)^{1/2}$.

If (5.1) is true, particularly if $\rho(t_1) \lesssim 1 - (I^*/n)^{1/2}$, the quantitative details of the second terms of (5.6) are not very important anyway. The duration of the transition and $E\{N_c(t)\}$ are dominated by the properties of the first term. If we neglect the second term of (5.6) for $z > 0$, $E\{N_c(t)\}$ can be approximated by

$$E\{N_c(t)\} \simeq [\text{Var}\{N^*(t)\}]^{1/2} H(E\{N^*(t)\}/[\text{Var}\{N^*(t)\}]^{1/2}) .$$

CHAPTER IV. EQUILIBRIUM DISTRIBUTIONS

1. _Introduction._ In Chapter III we described the behavior of queues when the expected service time was small, first compared with the duration of the rush hour, and second compared with the duration of any transitions during which queueing may or may not occur. There were many different types of behavior, particularly in the latter case, because new time constants associated with the relaxation time of fluctuations had to be compared with $E\{S\}$ and various time parameters of $\rho(t)$. So far, we have not done much more than to identify these various types, describe some qualitative properties, and suggest how they might be analysed.

Basically, the suggested procedure was that, when stochastic queueing lasted for a time large compared with $E\{S\}$, one should use diffusion approximations to describe the behavior of the customer queue $N_c(t) = N(t)$ if $N(t) > 0$, and use some normal distributions to approximate the distribution of the number of idle servers $N_s(t) = - N(t)$ if $N(t) < 0$; then try to piece the two together so as to obtain the complete distribution of $N(t)$.

There is little question that, if $N(t) > 0$ is almost certain, one can use diffusion approximations, or, if $N(t) < 0$ is almost certain that one can use normal approximations. In the former case, the behavior of $N(t)$ must be essentially the same as for a single server with mean service time $E\{S\}/n$, and, in the latter case, it will be like a service with infinitely many servers. As a practical matter, however, there are many situations in which neither of these idealizations is very accurate, and, in passing through transitions, one must switch from one to the other. It would, therefore, be useful if one could interpolate between these extremes and obtain some at least crude approximations when both $N(t) > 0$ and $N(t) < 0$ have a non-negligible probability.

The need of some interpolation, however, is not as great as the above might imply, otherwise we would have faced the problem sooner. If $\rho(t)$ is slowly varying with time, most of the types of behavior described in Chapter III existed already for the single server system and arise after $N(t) > 0$ has become almost certain. In the context of all the situations described in Chapter III, this interpolation is important in only a few cases or for a short period of time relative to the entire rush hour. One such situation is when $\rho(t)$ reaches a maximum value between $1 - (I*/n)^{1/2}$ and 1, in which case the distribution of $N(t)$ might approach an equilibrium distribution for which $N(t) < 0$ and $N(t) > 0$ are both important. In most other cases, however, the main purpose of the interpolation will be to estimate a correction to one of the two idealized behaviors during times when the behavior of the queue is of greatest practical importance (during transitions or times when $\rho(t)$ is near its maximum).

The logic behind the interpolation scheme is basically that discussed in Sections III-3 and 4. If $N(t_0) > 0$, the behavior of $N(t)$ for $t > t_0$ is not very sensitive to the history of $N(t)$ before time t_0, certainly not to the history prior to time about $t_0 - E\{S\}$, or the value of $E\{N(t_0)\}$, similarly if $N(t_0) < 0$. If, however, $N(t_0) > 0$ and $N(t_0) < 0$ are both likely possibilities, we are mainly concerned with values of $N(t_0)$ of order $(I*n)^{1/2}$ or $(I^{\dagger}n)^{1/2}$ and traffic intensities $\rho(t)$ with $|1 - \rho(t)| \lesssim (I*/n)^{1/2}$, for otherwise $N(t) < 0$ or $N(t) > 0$ will become virtually certain within a time of order $E\{S\}$ or was virtually certain within a time of order $E\{S\}$ before t_0; these cases have all been analyzed previously, mostly in Chapter II.

Suppose we let

$$F(z , t|z_0) = P\{N(t) < z|N(t_0) = z_0\} \tag{1.1}$$

be the distribution function of $N(t)$ given $N(t_0) = z_0$. Regardless of the value of z_0, this conditional distribution of $N(t)$ will be approximately normal if $t - t_0$ is of order $E\{S\}$ or less, but the properties of the mean and variance will depend upon z_0, particularly whether $z_0 > 0$ or $z_0 < 0$.

If $z_0 > 0$ and sufficiently large that $N(t) > 0$ is almost certain

$$E\{N(t)\} \simeq z_0 + [n/E\{S\}] \int_{t_0}^{t} [\rho(x) - 1]dx \tag{1.2}$$

as in (III 4.4). We will assume here, however, that $\rho(t)$ varies slowly enough so that for $t - t_0$ comparable with $E\{S\}$,

$$E\{N(t)\} \simeq z_0 + (t - t_0)[\rho(t_0) - 1]n/E\{S\} \quad \text{for} \quad z_0 > 0 . \tag{1.3}$$

On the other hand, if $z_0 < 0$, some servers are idle. Within a time comparable with $E\{S\}$, most customers in the service will have left and been replaced by new customers; $E\{N(t)\}$ will try to adjust to the equilibrium value of

$$E\{N(t)\} \simeq [\rho(t_0) - 1]n \quad \text{for} \quad z_0 < 0 .$$

Although $N(t)$ may become positive before this value is reached, the queueing will not affect the value of $E\{N(t)\}$ until a time of order $E\{S\}$ after the queueing occurs. Also, if the amount of queueing is small, it will have little effect. We can use (1.4) at least over a limited range of $t - t_0$ comparable with $E\{S\}$.

The two forms (1.3) and (1.4) are quite different; (1.3) depends upon z_0 and $t - t_0$, whereas (1.4) is independent of these. If $\rho(t_0) < 1$, the mean queue in (1.3) will decrease, eventually causing $N(t)$ to leave the region where (1.3) applies (i.e. $N(t)$ becomes negative), but (1.4) gives a negative value tending to preserve the condition $N(t) < 0$. It is possible to achieve an equilibrium. If, however, $\rho(t_0) > 1$, the expected queue (1.3) will increase; also $E\{N(t)\} > 0$ in (1.4). Eventually $N(t) > 0$ will become certain.

If $z_0 > 0$ and $N(\tau)$ stays positive until time t , $\text{Var}\{N(t)\}$ will be the sum of the variances of the arrivals and of the service completions between time t_0 and t . The former is, by hypothesis, approximately $I\lambda(t_0)(t - t_0)$, proportional to $(t - t_0)$. The service process, however, is the superposition of n renewal processes. Over times short compared with $E\{S\}$, this superposition looks like a Poisson process and gives a variance of approximately $(t - t_0)n/E\{S\}$ (equal to the mean number of service completions), but for $(t - t_0) \gg E\{S\}$, the variance is approximately $c_S^2(t - t_0)n/E\{S\}$. Actually the latter value is the more appropriate one for $(t - t_0)$ comparable with $E\{S\}$, but it may be helpful in some of the interpolations to recognize that for times short compared with $E\{S\}$, the "effective value" of c_S^2 may be closer to 1 than its true value. Thus

$$\text{Var}\{N(t)\} \simeq [I\rho(t) + c_S^2](t - t_0)n/E\{S\} . \tag{1.5}$$

In most of the following, however, $\rho(t)$ will be sufficiently close to 1 that we will want to simplify this to

$$\text{Var}\{N(t)\} \simeq [I + c_S^2](t - t_0)n/E\{S\} = I^\dagger(t - t_0)n/E\{S\} \qquad (1.5a)$$

If $z_0 < 0$, and $N(t)$ stays less than 0 , the system will behave as in Section II-2. After a time comparable with $E\{S\}$, $\text{Var}\{N(t)\}$ will approach a value

$$\text{Var}\{N(t)\} \simeq I*[E\{N(t_0)\} + n] \simeq I*\rho(t_0)n \simeq I*n . \qquad (1.6)$$

Whereas the variance in (1.5) increases proportional to $(t - t_0)$, (1.6) is nearly constant.

The process $N(t)$ is certainly not a Markov process, mainly because the service has a memory lasting for a time of order $E\{S\}$. This is, of course, the reason why any precise mathematical analysis of such a system is so complicated as to be of no practical value. In some respects, however, $N(t)$ does behave like a Markov process with normally distributed transition probabilities as described above over time intervals $t - t_0$ comparable with $E\{S\}$.

If $N(t)$ stays positive for times large compared with $E\{S\}$, the system really does behave like a Markov process. Furthermore, the change in $N(t)$ during short times is small, so that $F(z , t|z_0)$ satisfies a diffusion equation of the type.

$$\frac{E\{S\}}{n} \frac{\partial F}{\partial t} = [1 - \rho(t)] \frac{\partial F}{\partial z} + \frac{1}{2} \frac{\partial}{\partial z} [I\rho(t) + c_S^2] \frac{\partial F}{\partial z} . \qquad (1.7)$$

This equation, which can be derived from the above normally distributed transition probabilities[11], is valid for any values of z such that one can be sure that $N(t)$ has been positive for at least a time of order $E\{S\}$ prior to time t . If $\rho(t)$ is close to 1, this means for z larger than about $(I^\dagger n)^{1/2}$ and perhaps also $(I*n)^{1/2}$ (if these are significantly different). This is valid even if most of the probability lies below $(I^\dagger n)^{1/2}$, but is most useful if there is only a small probability for $N(t) \lesssim (I^\dagger n)^{1/2}$. This is, of course, the justification for some of the qualitative interpretations in Chapter III.

If $N(t)$ could stay negative for several mean service times, $N(t)$ would also behave approximately like a Markov process over time steps of about $E\{S\}$. The Markov process, however, reaches its equilibrium distribution in about one step.

For values of z_0 very close to 0 , one could imagine a Markov process with one set of transition probabilities for $z_0 > 0$ and another for $z_0 < 0$ but if these are not the same at $z_0 = 0_+$ and $z_0 = 0_-$, one would have reason to question the accuracy of approximating the behavior of $N(t)$ by a Markov process near the origin. Clearly, if $z_0 = 0_+$, $N(t)$ could become negative very soon after t_0 ; and, if $z_0 = 0_-$, it could become positive after t_0 . Thus at $z_0 = 0$, the evolution of $N(t)$ must be some compromise between the behaviors proposed for $z_0 > 0$ and $z_0 < 0$. If we use a Markov process with time steps of $E\{S\}$, the changes in the mean from $z_0 = 0$, $E\{N(t)\}$, are the same in (1.3) and (1.4), but the variances in (1.5) and (1.6) will not agree unless $I^* = I^+$ (which is the case for regular service $S = s$) .

We expect that the shape of the distribution of $N(t)$ near $N(t) = 0$ will depend upon the service distribution and that, in general, it will not be accurately represented by any simple formula. Although it would be interesting to know some qualitative features of the distribution near $N(t) = 0$, this is not of much practical concern. It is more important to know the behavior of the positive tail of the distribution of $N(t)$, and perhaps some moments of the distribution, particularly $E\{N_c(t)\}$, maybe also $E\{N_s(t)\}$, the expected queue length of customers and servers. Actually the distribution of $N(t)$ near 0 contributes very little to either $E\{N_c(t)\}$ or $E\{N_s(t)\}$; we would be willing to accept rather large errors in the distribution near 0 as long as they did not affect the estimates of the tail distribution.

Although we will find it convenient to think of $N(t)$ as being like a Markov process as described above, it is not simple to analyse the Markov process quantitatively. We would prefer some even simpler approximations. For example, it would be very convenient if we could find some simple "effective boundary conditions" near $z = 0$, which, when used in conjunction with (1.7), would accurately describe the evolution of the tail distribution.

As a means of obtaining some of the above limited objectives, we will also find equation (III 3.16) very useful. It is quite accurate, and it relates $E\{N_s(t)\}$ to $E\{N_c(t)\}$, neither of which is very sensitive to the distribution of $N(t)$ near

$N(t) = 0$.

Since we will be forced to invent some rather crude (and unconvincing) schemes of approximations, it will be helpful first to apply some of the above arguments to the evaluation of equilibrium distributions. We can then make some quantitative comparisons with known results for special systems, particularly the $M/M/n$ and $M/D/n$ systems.

2. <u>Approximate equilibrium distributions</u>. If $\rho(t) = \rho < 1$ is a constant, the distribution of $N(t)$ will approach an equilibrium distribution for sufficiently large t (or if $\rho(t)$ varies slowly enough, it will approximately follow the equilibrium distribution corresponding to the value of ρ evaluated at time t) .

If we let

$$F(z) \equiv P\{N(t) < z\} \tag{2.1}$$

be the distribution function for the equilibrium distribution, then be setting the time derivative equal to zero in (1.7), we conclude that

$$1 - F(z) \simeq A \exp [- z2(1 - \rho)/(I\rho + C_S^2)] \simeq A \exp [- z2(1 - \rho)/I^\dagger] \tag{2.2}$$

for some, as yet unknown, constant A . This is true at least for values of $z \gtrsim (I*n)^{1/2}$ or $(I^\dagger n)^{1/2}$, for which $N(t)$ will stay positive for at least a few service times.

If (2.2) were valid for all $z > 0$, we would conclude that the conditional mean queue given that it were positive, $E\{N(t)|N(t) > 0\}$, had a value $\frac{1}{2} I^\dagger/(1 - \rho)$. But if this were large compared with $(I*n)^{1/2}$ or $(I^\dagger n)^{1/2}$, i.e., if $1 - \rho << (I^\dagger/n)^{1/2}$ or $(I^\dagger/n)^{1/2}(I^\dagger/I*)^{1/2}$, we would expect most of the probability distribution to satisfy (2.2) with A close to 1 . The unconditional expected queue length $E\{N_c(t)\}$ would then be approximately $\frac{1}{2} I^\dagger/(1 - \rho)$, in agreement with the known "heavy traffic" behavior of any single or multiple server system. We are concerned here, however, mainly with the situations in which $1 - \rho$ is comparable with $(I^\dagger/n)^{1/2}$.

On the other hand, we also know that if $N(t)$ stays negative for a time comparable with $E\{S\}$ that the distribution of $N(t)$ would be approximately normal with mean $(\rho - 1)n$ and variance $I*\rho n$ in accordance with (1.4) and (1.6). If $N(t) \underset{\sim}{<} - (I*n)^{1/2}$, the conditional distribution of $N(t)$ must be approximately normal, and so

$$F(z) \simeq B\Phi([z + (1 - \rho)n]/(I*\rho n)^{1/2})\qquad(2.3)$$

for $z \underset{\sim}{<} - (I*n)^{1/2}$.

If $1 - \rho$ were larger than about $(I*/n)^{1/2}$, this normal distribution would barely reach $z = 0$. We would expect negligible queueing and $B = 1$. The behavior in this case would agree with that described in Section III-3.

Our problem now is to interpolate between the formulas (2.2) and (2.3) over the range in which z is of order $(I^{\dagger}n)^{1/2}$ or $(I*n)^{1/2}$, and to determine the constants A and B so that the formulas match properly. These interpolations should be "smooth" because, in this range of z , the behavior of $N(t)$ is, in some sense, a compromise between the behaviors for $N(t) > 0$ and $N(t) < 0$. One should not assume from this, however, that the extrapolation of (2.2) to $z \sim 0$ will smoothly join the extrapolation of (2.3) to $z \sim 0$, but it is, nevertheless, instructive to look at the behavior of 2.3) near $z = 0$.

The probability density associated with (2.3) reaches its maximum at $z = - (1 - \rho)n < 0$ and is decreasing with z at $z = 0$. At least for a short range of z near some point z_1 , it is also decreasing approximately like an exponential in the sense that (for $\rho \sim 1$)

$$\exp[- (z + (1 - \rho)n)^2/2I*n] = \exp[- (z - z_1 + (1 - \rho)n + z_1)^2/2I*n]$$
$$= \{\exp[- ((1 - \rho)n + z_1)^2/2I*n]\}\{\exp[- (z - z_1)^2/2I*n]\}\qquad(2.4)$$
$$\times \{\exp[- (z - z_1)((1 - \rho)n + z_1)/I*n]\} .$$

For sufficiently small $z - z_1$, we can replace the second factor of (2.4) by 1 ; it is the last factor that is of particular interest since the probability density for (2.2) is also an exponential in $z - z_1$, for all $z > 0$.

As z_1 varies from $- (1 - \rho)n$ to $+\infty$, the parameter in the exponential (last factor) of (2.4) will cover all values from 0 to ∞ . If we wanted to match (2.3) smoothly to (2.2), we could do so at a point z_1 chosen so that the exponential factor of (2.4) matched that of (2.2),

$$\frac{(1 - \rho)n + z_1}{I^* n} = \frac{2(1 - \rho)}{I^\dagger}$$

i.e.

$$z_1 = n(1 - \rho)(- 1 + 2I^*/I^\dagger) . \tag{2.5}$$

If we also choose A and B in (2.2) and (2.3) so that both the distribution function and the probability densities are continuous at z_1 , the derivative of the probability density will also be continuous at z_1 .

Although the above procedure seems intuitively very appealing, it is usually incorrect. It appears to be accurate in two cases. First, if $I^\dagger = 2I^*$, then $z_1 = 0$ in (2.5). This is where it would seem that we should join the formulas (2.2) and (2.3). This condition is realized for exponentially distributed service times for which $2I^* = I^\dagger = 1 + I$. Since exponential service has special Markov properties, it is not surprising that this is also special in (2.5). We shall later confirm that the above matching is very accurate for exponentially distributed service.

The second case in which this matching may be appropriate is if $1 - \rho \gtrsim (I^*/n)^{1/2}$ and $2I^* \geq I^\dagger$, although, in this case, it makes little difference anyway. If $1 - \rho \gtrsim (I^*/n)^{1/2}$, we have previously argued, as in Section III 3, that the distribution of $N(t)$ should be approximately normal for either positive or negative values of $N(t)$ because the queue would survive for only a short time. If $z_1 > 0$, there is little probability for $N(t) > z_1$ anyway. Perhaps what probability that is left beyond z_1 could be fitted better to an approximate exponential tail than a normal tail.

In all cases, however, it must be true that

$$E\{N_s(t)\} = n(1 - \rho) ,$$

as in (III 3.16). Since the value of $E\{N_s(t)\}$ is not very sensitive to errors in the distribution of $N(t)$ near 0 , we might use (2.3) for all $z < 0$ and obtain

$$n(1 - \rho) \simeq B(I^* \rho n)^{1/2} H((1 - \rho)(n/\rho I^*)^{1/2}) ,$$

or

$$B \simeq \kappa/H(\kappa) \qquad (2.6)$$

with

$$\kappa \equiv (1 - \rho)(n/\rho I*)^{1/2} .$$

It is interesting to observe that this value of B , and therefore (2.3)

$$F(z) \simeq [\kappa/H(\kappa)]\Phi(\kappa + z/(I*\rho n)^{1/2}) \quad , \quad z < 0 \qquad (2.3a)$$

do not depend on $I^\dagger$. Although, for $2I^\dagger \neq I*$, we can expect some distortion in this distribution as z becomes close to 0 , which would be expected to affect the probability density, these distortions should not have much effect upon the cumulative probability up to $z = 0$. Thus, we expect, generally,

$$P\{N(t) < 0\} = F(0) \simeq \kappa\Phi(\kappa)/H(\kappa)$$

and consequently also the total probability for a non-zero queue

$$P\{N(t) > 0\} = 1 - F(0) = 1 - \kappa\Phi(\kappa)/H(\kappa)$$
$$= [1 + (2\pi)^{1/2}\kappa\Phi(\kappa) \exp (\kappa^2/2)]^{-1} \qquad (2.7)$$

to be quite insensitive to $I^\dagger$.

This relation represents the key to the interpolation because it describes the partition of the probability between $N(t) < 0$ and $N(t) > 0$. The function (2.7) is shown in Fig. IV.1 along with several approximating curves.

For $\kappa \gg 1$,

$$P\{N(t) > 0\} \simeq (2\pi)^{-1/2}\kappa^{-1} \exp (- \kappa^2/2) + 0[\exp (- \kappa^2)] . \qquad (2.7a)$$

In Section III.3, we predicted that for $1 - \rho \gtrsim (I*/n)^{1/2}$, i.e., $\kappa \gtrsim 1$, the distribution of $N(t)$, including $N(t) > 0$, should be approximately normal with a variance $I*n$, independent of $I^\dagger$ (at least to a first approximation). According to this normal approximation, we should have

$$P\{N(t) > 0\} \simeq 1 - \Phi(\kappa) \quad \text{for} \quad \kappa \gtrsim 1 . \qquad (2.7b)$$

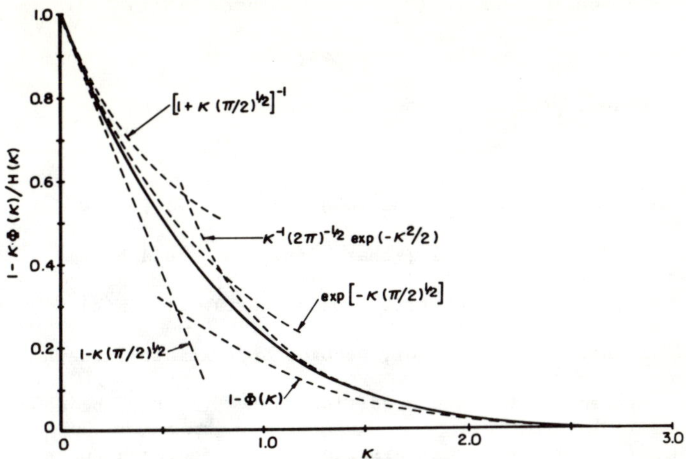

Fig. IV.1 - The solid line shows $1 - \kappa\Phi(\kappa)/H(\kappa)$, which
is an approximation for $P\{N(t) > 0\}$ eq. (2.7), κ , given
by (2.6) is proportional to $1 - \rho$. The broken line
curves represent various analytic approximations for
$\kappa \ll 1$ or $\kappa \gg 1$.

The asymptotic expansion of $\Phi(\kappa)$ for $\kappa \gg 1$ is

$$1 - \Phi(\kappa) \simeq (2\pi)^{-1/2}\kappa^{-1} \exp(-\kappa^2/2)[1 + 0(\kappa^{-1})] \quad \text{for} \quad \kappa \gg 1 .$$

This agrees with (2.7a) for $\kappa \gg 1$, but the asymptotic expansion is not accurate

for $\kappa \simeq 1$. The first term of (2.7a), and (2.7b) are both shown in Fig. IV.1 by

broken lines. For $\kappa \gtrsim 1$, both are quite close to (2.7), the former is slightly

high, the latter slightly low. The agreement between (2.7), (2.7a) and (2.7b) is

far better than simply an agreement among terms in an asymptotic expansion, however;

the agreement depends upon what types of functions one uses in the expansion.

For $\kappa \ll 1$, we can approximate (2.7) by

$$P\{N(t) > 0\} \simeq 1 - \kappa(\pi/2)^{1/2} \tag{2.7c}$$

or

$$\simeq [1 + \kappa(\pi/2)^{1/2}]^{-1} \tag{2.7d}$$

or

$$\simeq \exp[-\kappa(\pi/2)^{1/2}] . \tag{2.7e}$$

For small κ , these are all essentially equivalent. Which form one chooses is a matter of convenience. The form (2.7c) shows that $P\{N(t) < 0\}$ is approximately proportional to κ . Of the three formulas, the exponential form (2.7e), however, is most accurate and, for some purposes, also quite convenient. These approximations are also shown in Fig. IV.1.

We are mostly concerned with the behavior of (2.7) for $\kappa \lesssim 1$. We, of course, expected for $\kappa \to 0$, i.e. $\rho \to 1$, that $N(t) > 0$ would become almost certain (thus $F(0) \to 0$) . While the distribution of $N(t)$ for $N(t) < 0$ remains approximately normal in shape, and, in fact, becomes approximately one-half of a normal distribution centered at 0 , the condition that $E\{N_s(t)\}$ be equal to $n(1 - \rho)$ forces the amplitude of the normal distribution to decrease approximately proportional to $(1 - \rho)$.

Having obtained a suitable formula for $P\{N(t) > 0\}$, we must now postulate a shape for the distribution of $N(t) > 0$, so we can estimate $E\{N_c(t)\}$. There is not much question as to what we should do if $2I^* = I^\dagger$; we will simply join the exponential distribution (2.2) to (2.3) at $z = 0$. But even if $2I^* \neq I^\dagger$, we know that for $\kappa \lesssim 1/2$, say, at least half of the probability distribution has $N(t) > 0$ and a significant part of this probability must have come from realizations of $N(t)$ that have stayed positive for a time of order $E\{S\}$. The main effect of the "soft reflecting barrier" is that it captures some of the probability that would otherwise have been in an approximately exponential distribution like (2.2).

Of all the things, such as the probability density or the rate of decay of the density, etc., that we might wish to join smoothly at or near $z = 0$, it is most important that the distribution function join smoothly. As a first approximation at least, we might take

$$1 - F(z) \simeq [1 - \kappa\Phi(\kappa)/H(\kappa)] \exp[- z2(1 - \rho)/I^\dagger] \ , \quad z > 0 \tag{2.8}$$

and

$$\begin{aligned}
E\{N_c(t)\} \ &\simeq \ \int_0^\infty dz[1 - F(z)] = [1 - \kappa\Phi(\kappa)/H(\kappa)]I^\dagger/2(1 - \rho) \\
&= \ (nI^*)^{1/2}\kappa^{-1}[1 - \kappa\Phi(\kappa)/H(\kappa)](I^\dagger/2I^*) \ .
\end{aligned} \tag{2.9}$$

Since $P\{N(t) < 0\}$ is independent of $I^\dagger$, these formulas, to the extent that they apply even for $2I* \neq I^\dagger$, indicate that $E\{N_c(t)\}$, for fixed κ and $I*$, is proportional to $I^\dagger$. This is certainly true for $\kappa \to 0$, where the diffusion approximation gives $E\{N_c(t)\} \simeq I^\dagger/(1 - \rho)$, but it is probably rather insensitive to the value of κ (at least for κ less than maybe $1/2$ or 1).

For $\kappa \lesssim 1/2$, we substitute (2.7c) into (2.9) to obtain

$$E\{N_c(t)\} \simeq \frac{I^\dagger}{2(1 - \rho)} - \left(\frac{nI*\pi}{2\rho}\right)^{1/2}\left(\frac{I^\dagger}{2I*}\right) . \tag{2.9a}$$

The first term is the usual result for $\kappa \to 0$, $\rho \to 1$; the second term gives a constant correction of order $n^{1/2}$, independent of $1 - \rho$. The formula applies, of course, only if the second term is less than the first. In many applications, however, this correction term is not negligible.

If, again for $\kappa \lesssim 1$, we substitute (2.7e) into (2.8), we obtain, for $z > 0$

$$1 - F(z) \simeq \exp\left\{- \frac{2(1 - \rho)}{I^\dagger}\left[z + \left(\frac{nI*\pi}{2\rho}\right)^{1/2}\left(\frac{I^\dagger}{2I*}\right)\right]\right\} . \tag{2.8a}$$

This form is quite convenient, first because (2.7e) is a very good approximation for $\kappa \lesssim 1$, and secondly because it suggests that, for $z > 0$, we might replace the soft barrier by an "effective hard barrier" located at a value of z of

$$- \left(\frac{nI*\pi}{\rho}\right)^{1/2}\left(\frac{I^\dagger}{2I*}\right) , \tag{2.10}$$

the same number as appears in (2.9a). The standard deviation of the normal distribution is $(n\rho I*)^{1/2}$. This effective position is at $+(\pi/2)^{1/2}(I^\dagger/2I*\rho)$ standard deviations inside the barrier. Whether or not one can really justify thinking of the barrier in this way, it is certainly intuitively appealing and must at least give a correct order of magnitude estimate of the effects of the barrier.

The above formulas cannot be "derived"; they are little more than wishful speculations that must be tested by numerical comparison with more accurate results. We would expect them to be quite accurate for $I^\dagger = 2I*$ for all κ and for $I^\dagger \neq 2I*$ if κ is small enough (hopefully for a significant range of κ out to perhaps $1/2$ or 1). They cannot possibly be correct, however, for $I^\dagger \neq 2I*$ and

$\kappa \gg 1$ because, in this case, we expect most of the distribution of $N(t)$ for $N(t) > 0$ to originate from occasional fluctuations from negative values of $N(t)$; and we expect $N(t)$ to be approximately normally distributed even for $N(t) > 0$ with a distribution that does not depend upon $I^\dagger$. Thus in (2.9), for $\kappa \gg 1$, the correct formula should not have the factor $I^\dagger/2I*$. Over some range of κ , this factor must make a transition from $I^\dagger/2I*$ to 1 .

Even if κ is not large, the extrapolation of the exponential distribution back to $z = 0$ so as to meet the normal distribution for $z < 0$ will force a discontinuity in the probability density (the derivative of $F(z)$) at $z = 0$, except for $I^\dagger = 2I*$. Our justification for the above approximations was that the distortion in the distribution would be confined to a range $|N(t)| \lesssim (I*n)^{1/2}$. If there were a significant probability for $N(t) \gtrsim (I*n)^{1/2}$, these distortions would not significantly affect the continuity of the extrapolations of $F(z)$. Furthermore $E\{N_c(t)\}$ is not very sensitive to errors in $F(z)$ near $z = 0$. If, however, κ is so large that most of the probability for $N(t) > 0$ lies between 0 and $(I*n)^{1/2}$, the above arguments certainly fail.

We have seen in Section II 2 that the value of $I*$ has a rather complicated dependence upon the service distribution which is not, in general, the same as that of $I^\dagger$. In most applications we expect $I^\dagger$ and $I*$ to be of comparable size, both comparable with 1 (the methods of analysis here are not guaranteed to be reliable in extreme situations). For regular service $I^\dagger = I = I*$; for exponential service $I^\dagger = 2I* = I + 1$; and for Poisson arrivals $I* = 1$, $I^\dagger = 1 + C_S^2$. In all these cases we expect $I* < I^\dagger$, but we do not ordinarily expect $I^\dagger$ to be much larger than $2I*$. Perhaps the only interesting exception to $I^\dagger \geq I*$ is the case $C_S \ll 1$ and $I \ll 1$, in which case $I^\dagger = I + C_S^2$ but $I*$ is likely to be about $I + C_S(2\pi)^{-1/2}$.

In most of the above examples with $I^\dagger < 2I*$, the exponential distribution (2.2) will have a more rapid rate of decay than the normal distribution (2.3) at $z = 0$, i.e. in (2.4) we would want to join the distributions at a $z_1 > 0$. By imposing the condition (2.6), however, we have already imposed a discontinuity in the probability density near $z = 0$.

We could speculate on the correct shape of $F(z)$ near 0 and invent a formula that would satisfy all the desired conditions, but it would necessarily be rather artificial. We expect the relation (2.7) to be quite accurate; therefore, any distortions of $F(z)$ should preserve the same partition between $P\{N(t) > 0\}$ and $P\{N(t) < 0\}$. The discontinuity in the probability density should be smoothed out, but after this smoothing we expect the probability density to still decay at some rate near $z = 0$ similar to the normal distribution with parameter I^* . Perhaps for a range of z near $z = 0$, the rate of decay of $1 - F(z)$ should also behave like an exponential with an effective value of I equal to $I + 1$, corresponding to $c_S^2 = 1$ (because the service completion process acts like a Poisson process over short times). Finally, once the normal distribution has acquired a decay as fast as the exponential distribution (2.2), the decay will continue to follow (2.2).

We will not try to invent a formula that will do all these things. Any such formula would be rather cumbersome, of doubtful accuracy, and, except in very special circumstances, give only small changes from the formulas described above.

We now turn our attention to some special cases for which more accurate formulas can be derived, so that we can test the accuracy of some of these estimates.

3. <u>Equilibrium distributions for M/M/n</u>. The equilibrium distributions for the M/M/n system are well-known and even fairly simple.

If λ is the stationary arrival rate, $\rho = \lambda E\{S\}/n < 1$, and p_j represents the equilibrium probability for the system (queue plus service) to contain j customers, then the p_j are given by [11],

$$
p_j = \begin{cases} p_0 (n\rho)^j/j & \text{for } j \leq n & \text{(3.1a)} \\ \\ p_n \rho^{j-n} & \text{for } j \geq n \,. & \text{(3.1b)} \end{cases}
$$

The p_n in (3.1b) can be expressed in terms of p_0 (or vice-versa) through (3.1a), and then p_0 or p_n determined by the condition

$$
\sum_{j=0}^{\infty} p_j = 1 \,. \tag{3.2}
$$

That exact formulas for the p_j are somewhat awkward is due solely to the fact that the series (3.2) cannot be summed in a simple form.

For $n = \infty$, (3.1) gives the Poisson distribution

$$p_j = e^{-n\rho}(n\rho)^j/j! \quad \text{for} \quad n = \infty \, . \tag{3.3}$$

For finite n , p_j behaves like a Poisson distribution for $j \leq n$, but like a geometric distribution for $j > n$.

A more convenient representation of the p_j is given by the recursion formulas (from which (3.1) is derived)

$$\frac{p_j}{p_{j-1}} = \begin{cases} \rho(n/j) & \text{if} \quad j \leq n \tag{3.4a} \\ \rho & \text{if} \quad j \geq n \, . \tag{3.4b} \end{cases}$$

For an equilibrium distribution to exist, it is necessary that $\rho < 1$. From (3.4a) we see that $p_j/p_{j-1} > 1$ for $j/n < \rho < 1$, but that $p_n/p_{n-1} = \rho < 1$. Thus p_j is monotone increasing in j until j exceeds $n\rho$ and is then monotone decreasing until $j = n$. The important point is that for $\rho < 1$, the range $j \leq n$ over which (3.4a) applies includes the peak of the Poisson distribution.

For $j > n$, $p_j/p_{j-1} = p_n/p_{n-1} = \rho$; the geometric distribution (3.4b) takes the last ratio of the p's from the Poisson distribution and continues it for $j > n$, but the geometric distribution decays less rapidly than the Poisson distribution.

Although most of the results of sections 1 and 2 were derived from conjectures based upon the anticipated behavior of systems with $S = s$, most of the formulas of these sections are certain to be verified here with great precision for the M/M/n system (much better than for the M/D/n system). In section 2, it was predicted that the distribution of $N(t)$ given $N(t) < 0$ will be approximately the same as it would be in the absence of queueing, i.e., the $p_j/\Sigma_1^n p_k$ for $j \leq n$ should be independent of n . For the M/M/n system, this happens to be exactly correct (it will not generally be exact for non-exponential service). It was also assumed that the distribution of $N(t)$ for $N(t) > 0$ should be approximately exponential (geometric), except possibly for $0 < N(t) \lesssim (I*n)^{1/2}$. This is also exactly correct for the M/M/n system, for all $j > n$, although the geometric distribution (3.1b) does not have quite the same rate of decay as (2.2).

It has also been assumed that the distribution of $N(t)$ given $N(t) < 0$ would be approximately normal; but it is well-known that the Poisson distribution is approximately normal [5] for large values of its parameter $n\rho$ (but for reasons somewhat different from those used in Chapters I to III). This difference between the normal and the Poisson distribution is, basically, the main source of error in the approximations of Section 2 as applied to the $M/M/n$ system. That the exponential and geometric distributions have slightly different rates of decay is related to this, since in Section 2 we have joined the exponential distribution smoothly to the normal distribution instead of the Poisson distribution.

The accuracy of the approximations of Section 2 are illustrated in Fig. IV.2.

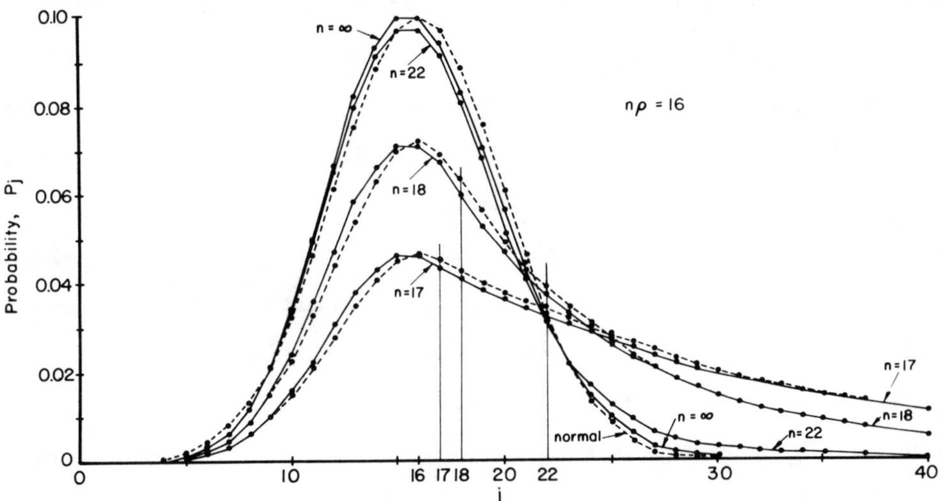

Fig. IV.2 – The probability p_j that there are j customers in an $M/M/n$ system, $(j - n)$ in the queue, for $n\rho = 16$. The solid line curve for $n = \infty$ is the Poisson distribution, the solid line curves for $n = 22, 18,$ and 17 are exact distributions. The broken lines are the approximate distributions, normal distribution with an exponential tail for $j > n$.

Instead of holding n fixed and varying λ , we have chosen to keep λ fixed at $\lambda/E\{S\} = n\rho = 16$ (the expected number of busy servers) and vary n , so that the shape of the Poisson distribution (3.1a) remains constant. Discrete values of the p_j are plotted vs. j with successive values of p_j , p_{j+1} joined by straight line segments. Only the values of the p_j at integer j are meaningful; the line segments are drawn only to assist in distinguishing the different curves and to suggest the similarity with a continuum of states.

The solid line curves are the exact distributions (3.1a, b) for $n = \infty$, n = 22, 18, and 17 ($\kappa = \infty$, 3/2, 1/2, and 1/4). The broken line curve labeled "normal" is the normal approximation to the Poisson distribution. It is essentially the probability density associated with (2.3) for $n \to \infty$ ($\kappa \to \infty$) . In (2.3), z measures the excess or deficiency of queueing customers or servers and is related to the j of Fig. IV.2 by $z = j - n$. Thus in (2.3), the normal distribution has a mean at $j = \rho n = 16$ and variance of $\rho n = 16$. The normal distribution and the Poisson distribution are rather close for $\rho n = 16$, the main error arising from the fact that the Poisson distribution is slightly skewed.

We have omitted the approximate curve for $n = 22$, but by comparing the exact curve for $n = \infty$ and $n = 22$, one can see that, whereas the two curves differ only by a constant factor for $j \leq 22$ and decay at about the same rate for a few values of j beyond 22 , the curve for $n = 22$ has a larger (geometric) tail than for $n = \infty$. This illustrates the fact that, for sufficiently large κ , the distribution of p_j , even for $j > n$ $(z > 0)$ is approximately normal, as discussed above and in Section III 3. The form of the errors, however, is such that the normal approximation at $\kappa \lesssim 3/2$ will not give a very accurate estimate of $E\{N_c(t)\}$ (the mean weights heavily the errors in the tail of the distribution).

For $n = 18$ and $n = 17$, the approximate curves from (2.3) and (2.8) are again shown by broken lines. These approximations continue to be very accurate as κ decreases. The figure illustrates, however, how, as κ decreases, more probability is pushed into the tail while the amplitude of the distribution for $j < n$ decreases. For $n = 17$, $\kappa = 1/4$, however, there is still about a quarter of the probability in the states $j \leq n$. The shape of the distribution is changing very rapidly with n . There will not be an equilibrium distribution for $n = 16$.

In comparing exact values of $E\{N_c(t)\}$ with those given by (2.9), we find that the discrepancy depends upon whether we take the effective value of $I^\dagger$ to be $1 + \rho$ as suggested by the diffusion equation (1.7) with $I = C_S^2 = 1$, or to be 2ρ as would be suggested by trying to fit an exponential distribution smoothly to a normal distribution with variance ρn (this is the value used in the evaluation of Fig. IV.2), or to be 2. The exact result lines somewhere between that which results from choosing $I^\dagger = 2\rho$ or 2. The methods used above are not sufficiently refined, however, to make it clear which value of $I^\dagger$ is most appropriate. We have indiscriminately replaced ρ by 1 in any factors involving variances. The fractional error in (2.9) is certainly no more than about $(1 - \rho)$.

4. __Equilibrium distributions for G/M/n__. The usual method of analysis of the M/M/n system is to recognize that it is a Markov process, and to derive the time-dependent rate equations for the state probabilities. To find the equilibrium distribution (3.1), one sets the time derivative equal to zero and solves the resulting equations for the $p_j[2]$.

One of the difficulties in studying the G/G/n system is that, in general, the future behavior of the system depends upon the past starting times of services presently in progress. This is not very important if the time constants such as the busy period or the time to go through a transition are large compared with $E\{S\}$, for then, as argued in Section 1, the system behaves approximately like a Markov process anyway and one can use diffusion approximations to estimate the queue lengths. If the service time is exponentially distributed, however, the past history of existing service times is completely irrelevant to the future and any non-Markov behavior must come from the arrival process.

We have been assuming, throughout this study, that, although the arrival process is not a Poisson process, statistical dependencies in arrivals persist for only a few interarrival times, a time which, for $n \gg 1$, is small compared with any other relevant time constants. For all practical purposes, any G/M/n system with an arrival process of this type will behave approximately as a Markov process. The distribution of $N(t)$ will therefore approximately satisfy a diffusion equation.

If, even for time dependent arrivals, we let

$$F(z , t) = P\{N(t) < z\} \tag{4.1}$$

be the distribution function of $N(t)$, then $F(z , t)$ will approximately satisfy an equation of the type [4]

$$\frac{\partial F(z , t)}{\partial t} = a(z , t) \frac{\partial F(z , t)}{\partial z} + \frac{1}{2} \frac{\partial}{\partial z} b(z , t) \frac{\partial}{\partial z} F(z , t) \tag{4.2}$$

in which $a(z , t)$ is the expected rate of decrease of $N(t)$ given that $N(t) = z$, and $b(z, t)$ is the rate of increase of the variance of $N(t)$ given $N(t) = z$.

The number of busy servers is n if $z > 0$ and $n + z$ if $z < 0$, therefore

$$
a(z, t) = \begin{cases} [n/E\{S\}][1 - \rho(t)] & \text{if } z > 0 \\ [n/E\{S\}][1 - \rho(t) + z/n] & \text{if } z < 0 \end{cases} \tag{4.3}
$$

and

$$
b(z, t) = \begin{cases} [n/E\{S\}][1 + I\rho(t)] & \text{if } z > 0 \\ [n/E\{S\}][1 + z/n + I\rho(t)] & \text{if } z < 0 \end{cases} \tag{4.4}
$$

For $z > 0$, this is, of course, equivalent to (1.7) with $c_S^2 = 1$. With exponentially distributed service times, however, the equation has been extended to $z < 0$

The boundary conditions to be satisfied by $F(z, t)$ are

$$
\begin{aligned}
F(- n, t) &\simeq F(- \infty, t) = 0 \\
F(+ \infty, t) &= 1 .
\end{aligned} \tag{4.5}
$$

For $n \gg 1$, and ρ comparable with 1, we do not expect $N(t)$ to reach $-n$ often enough to make any difference. For all practical purposes $(- n)$ is like $(- \infty)$.

For now, we are interested only in the equilibrium distributions. If we take $\rho(t) = \rho < 1$, and $F(z, t) = F(z)$ in (4.2), the solution of (4.2) gives for the probability density

$$
\frac{dF(z)}{dz} = [b(z)]^{-1} A' \exp\left[- 2 \int_0^z dx\, a(x)/b(x)\right]
$$

$$
= \begin{cases} A'' \exp[- 2z(1 - \rho)/(1 + \rho I)] & \text{for } z \geq 0 \tag{4.6a} \\ A'' e^{-2z} \left[\dfrac{1 + I\rho}{1 + I\rho + z/n}\right]^{2\rho n(1+I)-1} & \text{for } z \leq 0 \tag{4.6b} \end{cases}
$$

in which A'' is a constant to be determined so that $F(\infty) = 1$.

Equation (4.6a) gives the expected exponential distribution as in (2.2). For $I = 1$, this exponential agrees with the geometric distribution (3.1b) to within a fractional error of order $(1 - \rho)^2$ in the parameter of the exponential.

Despite the simple analytic form of (4.6b), this is not very convenient. For $I = 1$, it does not give the Poisson distribution (3.1a) exactly. For $\rho n \gg 1$, the first two asymptotic approximations to (4.6b) give

$$\frac{dF(z)}{dz} = B' \exp\left(- v^2/2\right) \exp\left[\frac{+ v^3}{6[(I + 1)n\rho/2]^{1/2}}\right] \quad \text{for } z \le 0 \qquad (4.6c)$$

with

$$v = \frac{z + n(1 - \rho) + 1/2}{[(I + 1)n\rho/2]^{1/2}}, \qquad (4.6d)$$

and B' a constant such that (4.6c) will equal (4.6a) at $z = 0$.

The first factor of (4.6c) represents a normal distribution with a mean at $- n(1 - \rho) - 1/2$ and a variance of $(I + 1)n\rho/2$. Since for exponential service $I^* = (I + 1)/2$, (4.6c) gives the same normal distribution as (2.3) except that the mean is shifted by $- 1/2$. This shift of the center is "small" compared with the standard deviation which is of order $n^{1/2}$ but, for moderately large n (10 or 20), would be noticeable. For $I = 1$, we see from Fig. IV.2 that the exact Poisson distribution does indeed have a maximum at about $1/2$ spacing less than the approximating normal distribution (2.3).

The second factor of (4.6c) skewes the distribution slightly. For v of order 1, the effects of this are of the same order of magnitude as the effects of the shift by $-1/2$ in the mode. One should not include one effect without the other because the shift by $1/2$ in the mean of the normal distribution is exactly balanced by a shift in the mean due to the skew factor; the mean of the distribution 4.6a is still $n(1 - \rho)$.

Despite the fact that (4.6b) does not give the Poisson distribution exactly for $I = 1$, it is considerably more accurate than the approximations discussed in the last section. If we were to draw a graph of (4.6) for $n\rho = 16$ as in Fig. IV.2, it would be so close to the exact distributions that one would hardly see the difference on such a scale as Fig. IV.2.

Our primary objective here was not to obtain improvements in the approximations of section 3. The solution (4.6) is more accurate than we need but also more complicated than we wish. From the accuracy of (4.6), however, we would conclude that to approximate a discrete Markov process by a diffusion process is <u>not</u> a major source of error in the context of the problems considered here.

If we had approximated $b(z , t)$ in (4.4) by its value at $z = -n(1 - \rho)$ i.e.,

$$b(z , t) \simeq [n/E\{S\}](I + 1)\rho \qquad \text{for} \quad z < 0 , \tag{4.4a}$$

the distribution (4.6) would have given the normal distribution (2.3). Thus the deviations from the normal distribution can be identified with the small change in $b(z , t)$ over the range of the distribution.

The value of $b^{1/2}$, in effect, determines the scale of length for the distribution. Thus in (4.6d), the effect of I is to change the scale of the normal distribution by a factor $[(I + 1)/2]^{1/2}$, without changing its mean. The exponential part of the distribution, however, is located near $z = 0$ where

$$b(z , t) = [n/E\{S\}][I\rho + 1] . \tag{4.4b}$$

This explains why (4.6a) contains the combination $I\rho + 1$ whereas (4.6d) contains the combination $I + 1$. Since we would not be concerned with the exponential part of the distribution unless ρ were quite close to 1 , we do not consider this effect to be very important. In section 3, for $I = 1$, this difference between $\rho + 1$ and 2 was usually disregarded.

In (4.6a), it appears that the effect of I is to change the scale of length by a factor of $(I\rho + 1)/2$ rather than $[(I + 1)/2]^{1/2}$ or $[(I\rho + 1)/2]^{1/2}$ as in (4.6d). But in (4.6d) we see also that a change in I would shift the center of the normal distribution relative to a rescaled z , or, in effect, also change the value of $n(1 - \rho)$. If we neglect the skew of the distribution or any other consequences of the z dependence of $b(z , t)$, the solution (4.6) depends upon only two parameters $[(I + 1)n\rho/2]$ and $n(1 - \rho)[(I + 1)n\rho/2]^{1/2}$, or, in effect, $[(I + 1)n/2]$ and $n(1 - \rho)[(I + 1)n/2]^{-1/2}$. If we were to change the scale of length z so that

$$z' = z \left(\frac{I + 1}{2}\right)^{-1/2} \tag{4.7}$$

and of $1 - \rho$ so that

$$1 - \rho' = (1 - \rho)\left[\frac{I + 1}{2}\right]^{-1/2} \quad , \tag{4.8}$$

the distribution (including the exponential part) with parameters ρ , n , and I as a function of z would map into the distribution with parameters ρ' , n , and I = 1 as a function of z' .

More generally, there are only three parameters in the exact distribution (4.6a,b). If one is willing to change n as well as ρ and the scale of z , it is possible to convert the distribution with $I \neq 1$ into an equivalent distribution with I = 1 . Specifically, if one substitutes z' = z

$$1 - \rho' = \frac{2(1 - \rho)}{2 + \rho(I - 1)} \quad , \qquad 1 - \rho = \frac{(1 - \rho')}{1 + \rho'\left(\frac{1 - I}{1 + I}\right)} \tag{4.9}$$

$$n' = n[1 + \rho(I - 1)/2] \quad , \qquad n = n'\left[1 + \rho'\left(\frac{1 - I}{1 + I}\right)\right] \quad ,$$

into (4.6), the distribution is mapped into the distribution associated with the parameters ρ' , n' with I = 1 .

Our conclusion is that properties of systems with $I \neq 1$, but exponential service, are essentially equivalent to suitably chosen M/M/n systems. Finally, one should notice that the reason why the shapes of the distributions of N are independent of I , except possibly for changes in scale and $n(1 - \rho)$, is that the variance coefficient that applies for N < 0 , the combination $(I + 1)/2$, is the same as the coefficient that applies for N > 0 . Since, as noted in Chapter III, the fluctuations in the arrivals and in the service interact differently for N < 0 than for N > 0 , we should not expect this invariance in shape necessarily to hold for service times that are not exponentially distributed.

5. <u>Equilibrium distributions for M/D/n or G/D/n</u>. Although most of the predictions of sections 1 and 2 and much of the theory of time-dependent behavior

discussed in Chapters I-III were based upon the properties of systems with regular service $S = s$, we saw that most of the conjectures of section 2 regarding equilibrium behavior were <u>exceptionally</u> accurate for systems with exponentially distributed service. Although sections 3 and 4 illustrated some of the properties of the approximate equilibrium distributions of section 2, they gave no indication of the typical accuracy of the approximations. The accuracy of the results of section 2 as applied to the $M/D/n$ or $G/D/n$ systems $(S = s)$ will certainly give a fairer indication of the accuracy of the methods than the results of sections 3 and 4.

Exact equilibrium distributions for the $M/D/n$ and $E_k/D/n$ systems (Erlang distributed interarrival times) have been derived and studied quite extensively [1, 12, 14]. The "exact" distributions, however, are usually given in the form of generating functions and involve roots of various transcendental equations. These forms are not very convenient for the analysis of systems with $n \gg 1$. Rather little attention has been given to asymptotic properties for large n .

The basic equation describing the evolution of queue lengths for the $G/D/n$ system is (I 2.11).

$$N(t) = [A_c(t) - A_c(t - s) - n] + \max [0 , N(t - s)] . \tag{5.1}$$

The first term represents the number of customer arrivals during time $(t - s , t)$, less n . It is assumed to be (nearly) statistically independent of the second term, the queue at time $t - s$. This equation relates the $N(t)$ at time t to its values at times $t - ks$ for integer k only. But if this has an equilibrium distribution on these discrete times, it also has the same equilibrium distribution on the continuous time t .

If, for $\rho(t) = \rho < 1$, we assume, as we have throughout, that the first term of (5.1) is normally distributed with mean $- n(1 - \rho)$ and variance $I*\rho n = I\rho n$ (for regular service), then it is convenient to consider a "dimensionless" form of (5.1). Let

$$N*(t) = N(t)/(I\rho n)^{1/2}$$
$$X(t) = [A_c(t) - A_c(t - s) - n]/(I\rho n)^{1/2} . \tag{5.2}$$

Then $N^*(t)$ satisfies

$$N^*(t) = X(t) + \max [0 , N^*(t - s)]$$

in which $X(t)$ is normal with variance 1 and mean

$$- n(1 - \rho)/(I\rho n)^{1/2} = - \kappa$$

as in (2.6).

If we let

$$F^*(z) = P\{N^*(t) < z\} \tag{5.3}$$

be the equilibrium distribution function of the $N^*(t)$, then $F^*(z)$ satisfies the Wiener-Hopf type integral equation

$$F^*(z) = \int_0^\infty dxF^*(x)(2\pi)^{-1/2}\exp[- \frac{1}{2} (z - x + \kappa)^2] . \tag{5.4}$$

Equations like (5.1) or (5.3) arise also in the analysis of a bulk service queue G/D/1 serving customers in batches of size n . Equation (5.4) was used previously [9] also in the analysis of a fixed-cycle traffic signal which, in essence, is a bulk service queue. Unfortunately, the solution of (5.4) cannot be represented in terms of elementary functions. One can derive various integral representations of the solution but, as yet, the solutions have not been tabulated.

One should first notice that, as a result of the transformation (5.2), there is only one parameter left in (5.4), κ . As in the case of the G/M/n system the dependence of the distribution upon I has been absorbed into the scale.

Fig. IV.3 shows some exact distributions of the p_j for the M/D/n system with $n\rho = 16$, drawn on the same scale as for the M/M/n system with $n\rho = 16$, as in Fig. IV.2. For $n = \infty$, the M/D/n and M/M/n systems both have a Poisson distribution with parameter 16, as does the M/G/∞ system for any service distribution (for any M/G/∞ system, $I^* = 1$). Thus the curves in Fig. IV.2 and Fig. IV.3 for $n = \infty$ are the same.

By comparing the curves for $n = 22$ ($\kappa = 3/2$) in Fig. IV.2 and Fig. IV.3 (relative to the common curve for $n = \infty$), one can see that, in Fig. IV.2, an exponential tail exists for $j \geq 22$ whereas in Fig. IV.3 the tail distribution for $j \geq 22$ still retains approximately the same Poisson (or normal) shape as for $n = \infty$. This

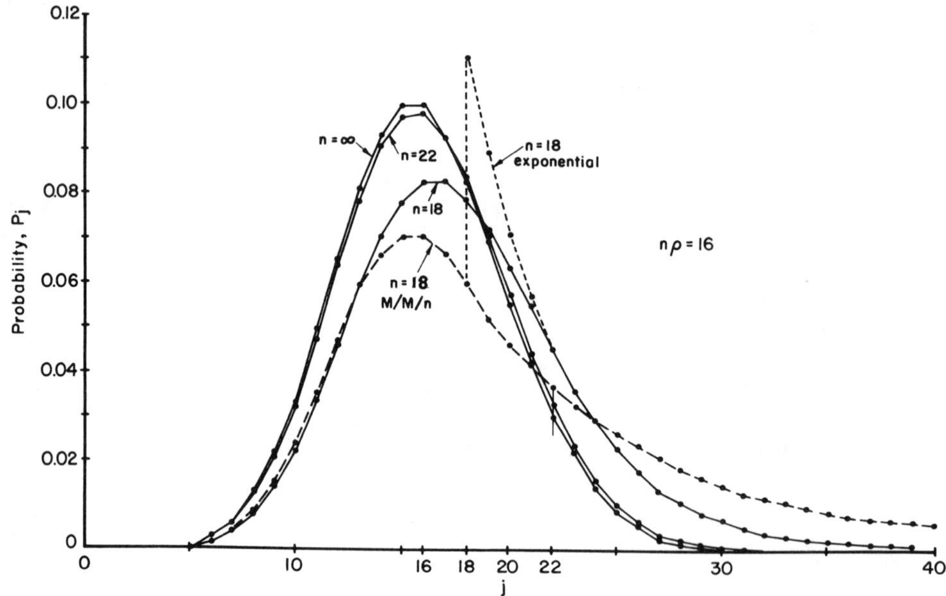

Fig. IV.3 - The probability p_j that there are j customers in an M/D/n system for nρ = 16 ; n = ∞ , 22, and 18. The broken line curve is for the corresponding M/M/n system; n = 18 .

agrees with the results of section III 3 where we predicted that, with S = s , the

distribution for N(t) should be approximately normal to a second approximation

(even for time-dependent arrivals), provided that $1 - \rho(t) \gtrsim (I/n)^{1/2}$. Equa-

tion III (3.4) and III (3.5) give for, $\rho(t) = \rho$, I = 1 ,

$$E\{N(t)\} \simeq - n(1 - \rho) + (\rho n)^{1/2}H(- \kappa) \tag{5.5}$$

$$Var\{N(t)\} \simeq n\rho[1 + H*(- \kappa)] . \tag{5.6}$$

For nρ = 16 , κ = 3/2(n = 22) , the second term of (5.5) shifts the mean by

about 0.1, and the second term of (5.6) increases the standard deviation by about

1%. Figs. IV.2 and IV.3 show that the net effect of these shifts (5.5) and (5.6)

for $j \lesssim 16$ is nearly the same for the M/D/n as for the M/M/n systems, despite

the differences in the formulas. This was to be expected, however, because, for

$j \lesssim 16$, N(t) has almost certainly been negative for at least one or two service

times prior to time t , and the behavior of the system should be almost as if n

were infinite. But for $17 < j \leq 22$, the p_j for n = 22 in Fig. IV.3 are higher

than for n = ∞ , whereas they are lower than for n = ∞ in the M/M/n system of

Fig. IV.2. In Fig. IV.2, the curve for n = 22 is a shift downward from n = ∞

over the range $j \leq 22$, whereas in Fig. IV.3 the shift is mostly to the right.

A normal distribution with mean and variance given by (5.5) and (5.6) will have approximately the same shape as the exact curve for $n = 22$ in Fig. IV.3, however, the 1% increase in the standard deviation from the second term of (5.6) will drop the peak of the normal distribution by only 1% relative to the peak for $n = \infty$. One can see in Fig. IV.3, that the peak actually drops by about 2%. The reason for this is that the second normal approximation is least accurate in the tails; it is designed to describe the shape mainly within 1 or 2 standard deviations of the mean. The effects of queueing push some probability into the tail, and the normal approximations (5.5), (5.6) tend to underestimate this even for moderately large κ (the correct distribution eventually decays exponentially). This error in turn causes an error in the amplitude of the normal part of the distribution.

The above qualitative effects, which are small and seem to be unimportant for $n = 22$, become amplified as κ decreases. Fig. IV.3 also shows the distribution of the p_j for $n = 18(\kappa = 1/2)$ in comparison with the corresponding curve (broken line) for the $M/M/n$ system with $n = 18$ taken from Fig. IV.2.

On the basis of the arguments given previously, one would not have expected a normal approximation to be very accurate for $\kappa = 1/2$ (it is supposed to apply for "large" κ), but (5.5) and (5.6) predict a displacement of the mean by about 0.8 and an increase in the standard deviation of about 9% compared with $n = \infty$. The peak amplitude of the distribution for $n = 18$ is actually about 17% less than for $n = \infty$, instead of 9%, but, except for this, the normal approximation would fit very well over the range of j from about 12 to 22. There is no question that if one joined an exponential distribution of the form (2.8) smoothly to the normal distribution satisfying (5.5) and (5.6), it would describe the distribution of the p_j very accurately. We will not pursue this point further, however, because this is a scheme which would apply specifically to the $M/D/n$ or $G/D/n$ systems but not more general types of service distributions. Also, this scheme would eventually fail even for the $M/D/n$ system for sufficiently small κ .

Our primary objective here is to illustrate some of the cruder methods of approximation proposed in section 2. It was suggested in (2.3a) that the distribution for $z < 0$ ($j < 18$) would follow approximately the same distribution as for

$n = \infty$ except for a constant decrease in amplitude, adjusted so as to give the cor-

rect expected number of idle servers. In the present situation with Poisson ar-

rivals, this implies that the p_j for the M/D/n system should be nearly equal to

those of the M/M/n system with the same ρ and n , except possibly for $N(t)$

near 0 , i.e., j near 18.

Fig. IV.3 does indeed show that the p_j for the M/D/n and M/M/n system are

nearly equal for $j \lesssim 14$, but for $14 < j \leq 17$, the p_j for the M/D/n become

appreciably larger than for the M/M/n . These differences are not enough, however,

to cause much change in the distribution function at $j = 17$, i.e. $F(-1) = \Sigma_1^{17} p_j$.

The difference in $F(-1)$ between the two systems is only about 0.03, compared with

a value of $F(-1)$ of nearly 0.5.

In (2.8) it was further proposed that we could approximate the distribution

function for $z > 0$ by an exponential distribution with an amplitude corrected for

the probability that $N(t) < 0$. The broken line curve labeled "n = 18 exponential"

is obtained from equation (2.8). As expected, this is not very accurate for short

queues, $j = 19$ or 20 ($0 < z < 1/2$) , but for $j > 21$ it is extremely accurate

(that the amplitude of the exponential distribution matches nearly perfectly the

exponential tail of the exact distribution must however be, in part, accidental; the

methods are not expected to be this accurate).

In essence, the methods of section 2 would approximate the distribution of the

p_j for the M/D/n system by those for the M/M/n (or a normal approximation to

these) for $j < n$, i.e., $j \leq 17$ in Fig. IV.3 for n = 18 , and then jump to the

exponential distribution for $j > n$ (because of the discontinuity in the approxima-

tions from either side, the value at $j = 18$ is left unspecified). Although this

appears to be a rather crude approximation, the largest errors occur near $N(t) = 0$

where they have little effect upon the moments $E\{N_c(t)\}$ or $E\{N_s(t)\}$.

The distributions of Fig. IV.3 were evaluated "exactly" from (5.1) in order to

compare them with the corresponding exact distributions for the M/M/n system (they

were evaluated by iteration of (5.1) with a Poisson distribution of arrivals in each

period s , with the aid of a K - E analogue computer, otherwise known as a slide

rule). To further compare the properties of the G/D/n system with the approximations of section 2, however, we shall use (5.4). This equation should be very
accurate for n >> 1 , certainly as compared with the approximations of section 2;
it is also independent of the I associated with the arrival process except for the
change of scale incorporated in the κ . The following results were deduced from
various integral representations of solutions of (5.4) and numerical integrations,
by methods which are too awkward to describe here.

In Fig. IV.4, the solid line curve labeled M/M/n is the function (2.7), shown

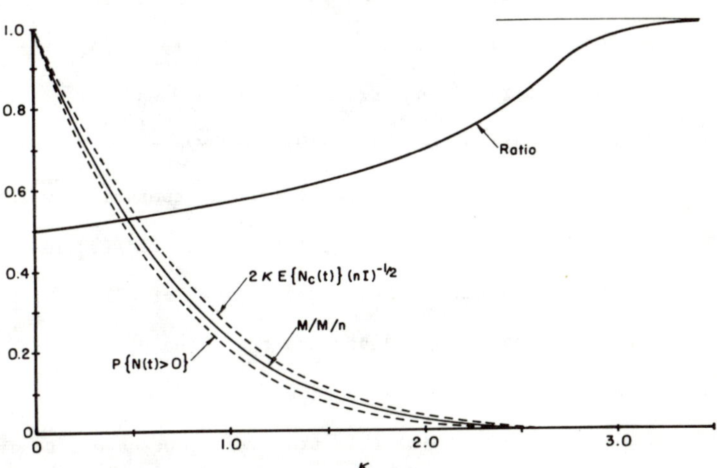

Fig. IV.4 - The curve labeled M/M/n represents both
P{N(t) > 0} and $\kappa E\{N_c(t)\}(nI*)^{-1/2}2I*/I^{\dagger}$ for the M/M/n
system, as a function of κ . The two broken line curves
are the corresponding quantities for the M/D/n system. The
curve labeled "ratio" is the ratio of $E\{N_c(t)\}$ for the
M/D/n to that of the M/M/n systems.

also in Fig. IV.1. For the M/M/n, or more generally for the G/M/n system, it
represents the normal approximation for P{N(t) > 0} as described by (2.7). According to (2.9), however, it also describes the expected queue length in the sense
that, for $2I* = I^{\dagger}$, it represents the function

$$(nI*)^{-1/2}\kappa E\{N_c(t)\}2I*/I^{\dagger} .\tag{5.7}$$

We saw in section 3 and 4 that this curve describes both P{N(t) > 0} and $E\{N_c(t)\}$
very accurately if the service time is exponentially distributed.

The curve of Fig. IV.4 labeled P{N(t) > 0} is the value of 1 - F(0) evaluated from the solution of (5.4). It was conjectured in section 2 that equation (2.7)

for P{N(t) > 0} would be insensitive to the arrival and service distributions except for the dependence upon I* which is absorbed as a scale parameter of κ , i.e. the curves labeled M/M/n and P{N(t) > 0} should be similar. The fractional difference between these curves becomes appreciable for large κ (where the value of P{N(t) > 0} is not likely to be very important anyway); otherwise the agreement is much better than one might have expected. At $\kappa = 1/2$, the difference between the two curves is about 0.03, which checks with the estimates from Fig. IV.3 based upon the Poisson arrival distributions with $n\rho = 16$.

The curve of Fig. IV.4 labeled $2\kappa E\{N_c(t)\}(nI)^{-1/2}$ is the value of (5.7) evaluated from numerical solutions of (5.4). According to (2.9), this curve should also be insensitive to the arrival and service distributions at least for small κ , i.e. it should also be similar to the curve labeled "M/M/n" . There was some question, however, as to how accurate this would be for κ comparable with 1, in view of the fact that, for sufficiently large κ , the factor $2I*/I^\dagger$ in (5.7) should be replaced by 1. This transition in the "effective value" of $2I*/I$ from 2 to 1 does not show very well on the graph of (5.7) because the value of (5.7) is quite small by the time κ is large enough for this to occur.

In order to amplify the effects of this transition, we have also drawn on Fig. IV.4 a graph labeled "ratio" which represents the ratio of the expected queue $E\{N_c(t)\}$ for the M/D/n to the value of $E\{N_c(t)\}$ for the corresponding M/M/n system, as a function of κ (this is 1/2 the value given by the curve "$2\kappa E\{N_c(t)\}(nI)^{-1/2}$" divided by the value from the curve "M/M/n"). For small κ , this ratio should be equal to the ratio of the factors $I^\dagger/2I*$, namely 1/2 ($I^\dagger =$ I* = 1 for the M/D/n ; $I^\dagger = 2I* = 1$ for the M/M/n), i.e., for heavy traffic the queue for the M/M/n system is twice as large as for the M/D/n . For large κ , however, when the queue is caused by the tail of a normal distribution (the same distribution for both systems), this ratio should go to 1. Fig. IV.4 shows that this transition from 1/2 to 1 is rather sharp and does not occur until κ is in the range of 2 to 3.

It is rather surprising that (2.9) should apply reasonably well out to $\kappa \simeq 2$. This suggests that the rather intuitive methods of section 2 will determine the equilibrium mean queue length quite well over most of the range of κ where it is of any importance, but, on the other hand, it suggests that the approximation of the complete distribution of $N(t)$ by a normal distribution for "large κ" may not give a very accurate estimate of the mean queue length except for $\kappa \gtrsim 2$ or 3. The use of a normal approximation for the distribution of $N(t)$ has been used extensively in previous sections, particularly for time-dependent queue behavior, in Chapter II and in section III 3. We also expected, at various times, that this might be a reasonable approximation if the probability in the tail $N(t) > 0$ was "small." One would ordinarily have expected that to go out one or two standard deviations on a normal distribution would be sufficient to guarantee a "small" tail probability but in the present calculations this is apparently not quite enough.

The above serves as a warning that one must be careful where and how one uses normal approximations for $N_c(t)$. The normal approximation actually gives a fairly good estimate of $E\{N_c(t)\}$ for the $M/D/n$ system; it is the $M/M/n$ system for which it is rather poor. (We saw already in Fig. IV.2 that the distortions from a normal distribution were quite large for $N(t) > 0$ even for $\kappa = 3/2$). Most of the previous applications of the normal distribution for $N(t)$, however, particularly in Chapter II, have been for time-dependent queues generated by a $\rho(t)$ which was increasing with time. It takes several service times, at least, for the tails of the equilibrium distributions to form. If the system passes through a transition in a few service times or less, one does not expect such large deviations from a normal distribution as found for the equilibrium distributions.

The above confirmation of the methods described in section 2 does not prove their general validity. It would be desirable to test them on still other systems, but (unlike the case of exponential service) there was no a-priori reason why the methods should have worked especially well for the $G/D/n$ system. We would expect the methods to apply at least this well for a wide range of service distributions "intermediate" between the exponential and the deterministic (for example, the Erlang distributions). It is not known, however, how well they will apply for service distributions with $C_S > 1$. Certainly the methods should provide a useful tool for quick estimates. They also describe, at least qualitatively, how fluctuations in the arrivals and the service interact, through the values of the parameters I^* and $I^\dagger$.

6. <u>Concluding comments</u>. When the study described here was first begun, the objective was to make a systematic study of the class of all n-channel service systems with $n \gg 1$, hopefully including various strategies for bulk service (the bus dispatching strategies). Fatigue from writing set in long before that goal was reached; the reader who has come this far must also be exhausted. To achieve the original goal would have required a treatise at least twice as long as the present one; the details becoming less and less interesting at each step. If this is ever achieved, it will be done by someone with more endurance than the author.

A more modest goal of just completing the analysis of systems which can serve only one customer at a time, which would require perhaps only one more chapter, has also been abandoned; not because it is difficult, but because it is straightforward, tedious, and unlikely to produce any surprises. The final unwritten chapter would have carried through, in more detail, the analysis of various transition behaviors described qualitatively in sections III.4 and IV.1, extended to non-equilibrium queues what was done above for the equilibrium distributions.

Such an extension would, for the most part, be nothing more than a correction for a "soft barrier" of the results obtained previously [10] for a "hard barrier." The practical purpose of such an analysis would be, primarily, to estimate queues when, during the evolution of the process, the queues are at or near their maximum size. As with the equilibrium distributions, one would expect the main effect of the soft barrier to be that it captures some probability in states with $N(t) < 0$ without otherwise affecting very much the shape of the conditional distribution given $N(t) > 0$. One would further expect that, in many cases, the soft barrier could be replaced by an effective hard barrier at a position approximately as given by (2.10).

The intended final phase of this study was simply to take the results obtained previously for the hard barrier and translate the distribution as if there were a hard barrier located as in (2.10). The accuracy of this would then be tested either by some further analytic estimates or by comparison with selected numerical simulations.

The previous analysis of the hard barrier was itself rather long because there are several different types of transition behavior. This final step would, therefore, involve rather extensive review, comparisons, etc. The only purpose would be to classify the various types of behaviors and to test the accuracy of something which seems certain to be more accurate than necessary in practical applications.

The _real_ purpose in attempting the present study, however, was not to make a catalogue of solutions; it was to illustrate some methods of approximation and demonstrate that it is possible to obtain approximate solutions, of fairly well-defined accuracy, for practical queueing problems. The queueing problems considered here are completely outside the limits of usefulness for conventional "exact" methods of queueing theory. This class of problems is itself rather narrow within the class of typical real life queueing problems, but it is the author's conviction that practical solutions can be found for nearly any practical problem if one concentrates attention on the things that are important and disregards the minute details.

In most practical queueing problems the most important issues of analysis are likely to concern the accuracy of possible mathematical models, rather than the accuracy of mathematical evaluations from the model. Although we have quite freely made use of postulates regarding normality of the distributions of cumulative counts, independence of counts in sufficiently long non-overlapping time intervals, etc., it is not obvious that these postulates will be true in the real world, no matter how plausible they may seem on theoretical grounds. For non-stationary arrivals, the only possible interpretation of a "probability distribution" is the distribution obtained from repetition of the observations "under identical conditions," which usually must be interpreted as on another day. Unfortunately such probability distributions are seldom what theoreticians would like them to be.

Rather than continue the purely theoretical game of deriving formulas for more and more hypothetical problems the next logical step would be to see if these results can be put to practical use. This will, in any particular problem, also be a rather tedious exercise which, hopefully, will be carried out before very long.

REFERENCES

1. Cooper, R. B., _Introduction to Queueing Theory_, Macmillan, New York (1972).

2. Cox, D. R., and Smith, W. L., _Queues_, Methuen, London (1961).

3. Cox, D. R., _Renewal Theory_, Methuen, London (1962).

4. Cox, D. R., and Miller, H. D., _The Theory of Stochastic Processes_, John Wiley, New York (1965).

5. Feller, W., _An Introduction to Probability Theory and Its Applications_, Vol. 1, 2nd ed. John Wiley (1957).

6. Haji, R., and Newell, G. F., Variance of the Number of Customers in an Infinite Channel Server, University of California, 1971. (unpublished)

7. Hurdle, V. F., Minimum Cost Schedules for a Public Transportation Route I, Theory II Examples, _Transportation Science_ 7, (1973)

8. Little, J. D. C., A Proof of the Queueing Formula. $L = \lambda W$, _Opns. Res._ 9, 383-387 (1961).

9. Newell, G. F., Approximate Methods for Queues with Application to the Fixed-Cycle Traffic Light, SIAM Rev. 7, 223-240 (1965).

10. Newell, G. F., Queues with Time-Dependent Arrival Rates, I, II, III, _J. Appl. Prob._ 5, 436-451, 579-590, 591-606 (1968).

11. Newell, G. F., _Applications of Queueing Theory_, Chapman-Hall, London (1971).

12. Riordan, J., _Stochastic Service Systems_, John Wiley, New York (1962).

13. Saaty, T. L., _Elements of Queueuing Theory with Applications_, McGraw-Hill (1961).

14. Syski, R. _Introduction to Congestion Theory in Telephone Systems_, Oliver and Boyd, London (1960).

Lecture Notes in Economics and Mathematical Systems

(Vol. 1–15: Lecture Notes in Operations Research and Mathematical Economics, Vol. 16–59: Lecture Notes in Operations Research and Mathematical Systems)

Vol. 1: H. Bühlmann, H. Loeffel, E. Nievergelt, Einführung in die Theorie und Praxis der Entscheidung bei Unsicherheit. 2. Auflage, IV, 125 Seiten 4°. 1969. DM 16,-

Vol. 2: U. N. Bhat, A Study of the Queueing Systems M/G/1 and GI/M/1. VIII, 78 pages. 4°. 1968. DM 16,-

Vol. 3: A. Strauss, An Introduction to Optimal Control Theory. VI, 153 pages. 4°. 1968. DM 16,-

Vol. 4: Branch and Bound: Eine Einführung. 2., geänderte Auflage. Herausgegeben von F. Weinberg. VII, 174 Seiten. 4°. 1972. DM 18,-

Vol. 5: Hyvärinen, Information Theory for Systems Engineers. VIII, 205 pages. 4°. 1968. DM 16,-

Vol. 6: H. P. Künzi, O. Müller, E. Nievergelt, Einführungskursus in die dynamische Programmierung. IV, 103 Seiten. 4°. 1968. DM 16,-

Vol. 7: W. Popp, Einführung in die Theorie der Lagerhaltung. VI, 173 Seiten. 4°. 1968. DM 16,-

Vol. 8: J. Teghem, J. Loris-Teghem, J. P. Lambotte, Modèles d'Attente M/G/1 et GI/M/1 à Arrivées et Services en Groupes. IV, 53 pages. 4°. 1969. DM 16,-

Vol. 9: E. Schultze, Einführung in die mathematischen Grundlagen der Informationstheorie. VI, 116 Seiten. 4°. 1969. DM 16,-

Vol. 10: D. Hochstädter, Stochastische Lagerhaltungsmodelle. VI, 269 Seiten. 4°. 1969. DM 18,-

Vol. 11/12: Mathematical Systems Theory and Economics. Edited by H. W. Kuhn and G. P. Szegö. VIII, IV, 486 pages. 4°. 1969. DM 34,-

Vol. 13: Heuristische Planungsmethoden. Herausgegeben von F. Weinberg und C. A. Zehnder. II, 93 Seiten. 4°. 1969. DM 16,-

Vol. 14: Computing Methods in Optimization Problems. Edited by A. V. Balakrishnan. V, 191 pages. 4°. 1969. DM 16,-

Vol. 15: Economic Models, Estimation and Risk Programming: Essays in Honor of Gerhard Tintner. Edited by K. A. Fox, G. V. L. Narasimham and J. K. Sengupta. VIII, 461 pages. 4°. 1969. DM 24,-

Vol. 16: H. P. Künzi und W. Oettli, Nichtlineare Optimierung: Neuere Verfahren, Bibliographie. IV, 180 Seiten. 4°. 1969. DM 16,-

Vol. 17: H. Bauer und K. Neumann, Berechnung optimaler Steuerungen, Maximumprinzip und dynamische Optimierung. VIII, 188 Seiten. 4°. 1969. DM 16,-

Vol. 18: M. Wolff, Optimale Instandhaltungspolitiken in einfachen Systemen. V, 143 Seiten. 4°. 1970. DM 16,-

Vol. 19: L. Hyvärinen, Mathematical Modeling for Industrial Processes. VI, 122 pages. 4°. 1970. DM 16,-

Vol. 20: G. Uebe, Optimale Fahrpläne. IX, 161 Seiten. 4°. 1970. DM 16,-

Vol. 21: Th. Liebling, Graphentheorie in Planungs- und Tourenproblemen am Beispiel des städtischen Straßendienstes. IX, 118 Seiten. 4°. 1970. DM 16,-

Vol. 22: W. Eichhorn, Theorie der homogenen Produktionsfunktion. VIII, 119 Seiten. 4°. 1970. DM 16,-

Vol. 23: A. Ghosal, Some Aspects of Queueing and Storage Systems. IV, 93 pages. 4°. 1970. DM 16,-

Vol. 24: Feichtinger, Lernprozesse in stochastischen Automaten. V, 66 Seiten. 4°. 1970. DM 16,-

Vol. 25: R. Henn und O. Opitz, Konsum- und Produktionstheorie. I. II, 124 Seiten. 4°. 1970. DM 16,-

Vol. 26: D. Hochstädter und G. Uebe, Ökonometrische Methoden. XII, 250 Seiten. 4°. 1970. DM 18,-

Vol. 27: I. H. Mufti, Computational Methods in Optimal Control Problems. IV, 45 pages. 4°. 1970. DM 16,-

Vol. 28: Theoretical Approaches to Non-Numerical Problem Solving. Edited by R. B. Banerji and M. D. Mesarovic. VI, 466 pages. 4°. 1970. DM 24,-

Vol. 29: S. E. Elmaghraby, Some Network Models in Management Science. III, 177 pages. 4°. 1970. DM 16,-

Vol. 30: H. Noltemeier, Sensitivitätsanalyse bei diskreten linearen Optimierungsproblemen. VI, 102 Seiten. 4°. 1970. DM 16,-

Vol. 31: M. Kühlmeyer, Die nichtzentrale t-Verteilung. II, 106 Seiten. 4°. 1970. DM 16,-

Vol. 32: F. Bartholomes und G. Hotz, Homomorphismen und Reduktionen linearer Sprachen. XII, 143 Seiten. 4°. 1970. DM 16,-

Vol. 33: K. Hinderer, Foundations of Non-stationary Dynamic Programming with Discrete Time Parameter. VI, 160 pages. 4°. 1970. DM 16,-

Vol. 34: H. Störmer, Semi-Markoff-Prozesse mit endlich vielen Zuständen. Theorie und Anwendungen. VII, 128 Seiten. 4°. 1970. DM 16,-

Vol. 35: F. Ferschl, Markovketten. VI, 168 Seiten. 4°. 1970. DM 16,-

Vol. 36: M. P. J. Magill, On a General Economic Theory of Motion. VI, 95 pages. 4°. 1970. DM 16,-

Vol. 37: H. Müller-Merbach, On Round-Off Errors in Linear Programming. VI, 48 pages. 4°. 1970. DM 16,-

Vol. 38: Statistische Methoden I, herausgegeben von E. Walter. VIII, 338 Seiten. 4°. 1970. DM 22,-

Vol. 39: Statistische Methoden II, herausgegeben von E. Walter. IV, 155 Seiten. 4°. 1970. DM 16,-

Vol. 40: H. Drygas, The Coordinate-Free Approach to Gauss-Markov Estimation. VIII, 113 pages. 4°. 1970. DM 16,-

Vol. 41: U. Ueing, Zwei Lösungsmethoden für nichtkonvexe Programmierungsprobleme. VI, 92 Seiten. 4°. 1971. DM 16,-

Vol. 42: A. V. Balakrishnan, Introduction to Optimization Theory in a Hilbert Space. IV, 153 pages. 4°. 1971. DM 16,-

Vol. 43: J. A. Morales, Bayesian Full Information Structural Analysis. VI, 154 pages. 4°. 1971. DM 16,-

Vol. 44: G. Feichtinger, Stochastische Modelle demographischer Prozesse. XIII, 404 Seiten. 4°. 1971. DM 28,-

Vol. 45: K. Wendler, Hauptaustauschschritte (Principal Pivoting). II, 64 Seiten. 4°. 1971. DM 16,-

Vol. 46: C. Boucher, Leçons sur la théorie des automates mathématiques. VIII, 193 pages. 4°. 1971. DM 18,-

Vol. 47: H. A. Nour Eldin, Optimierung linearer Regelsysteme mit quadratischer Zielfunktion. VIII, 163 Seiten. 4°. 1971. DM 16,-

Vol. 48: M. Constam, Fortran für Anfänger. VI, 143 Seiten. 4°. 1971. DM 16,-

Vol. 49: Ch. Schneeweiß, Regelungstechnische stochastische Optimierungsverfahren. XI, 254 Seiten. 4°. 1971. DM 22,-

Vol. 50: Unternehmensforschung Heute – Übersichtsvorträge der Züricher Tagung von SVOR und DGU, September 1970. Herausgegeben von M. Beckmann. VI, 133 Seiten. 4°. 1971. DM 16,-

Vol. 51: Digitale Simulation. Herausgegeben von K. Bauknecht und W. Nef. IV, 207 Seiten. 4°. 1971. DM 18,-

Vol. 52: Invariant Imbedding. Proceedings of the Summer Workshop on Invariant Imbedding Held at the University of Southern California, June – August 1970. Edited by R. E. Bellman and E. D. Denman. IV, 148 pages. 4°. 1971. DM 16,-

Vol. 53: J. Rosenmüller, Kooperative Spiele und Märkte. IV, 152 Seiten. 4°. 1971. DM 16,-

Vol. 54: C. C. von Weizsäcker, Steady State Capital Theory. III, 102 pages. 4°. 1971. DM 16,-

Vol. 55: P. A. V. B. Swamy, Statistical Inference in Random Coefficient Regression Models. VIII, 209 pages. 4°. 1971. DM 20,-

Vol. 56: Mohamed A. El-Hodiri, Constrained Extrema. Introduction to the Differentiable Case with Economic Applications. III, 130 pages. 4°. 1971. DM 16,-

Vol. 57: E. Freund, Zeitvariable Mehrgrößensysteme. VII, 160 Seiten. 4°. 1971. DM 18,-

Vol. 58: P. B. Hagelschuer, Theorie der linearen Dekomposition. VII, 191 Seiten. 4°. 1971. DM 18,-

Vol. 59: J. A. Hanson, Growth in Open Economics. IV, 127 pages. 4°. 1971. DM 16,-

Vol. 60: H. Hauptmann, Schätz- und Kontrolltheorie in stetigen dynamischen Wirtschaftsmodellen. V, 104 Seiten. 4°. 1971. DM 16,-

Vol. 61: K. H. F. Meyer, Wartesysteme mit variabler Bearbeitungsrate. VII, 314 Seiten. 4°. 1971. DM 24,-

Vol. 62: W. Krelle u. G. Gabisch unter Mitarbeit von J. Burgermeister, Wachstumstheorie. VII, 223 Seiten. 4°. 1972. DM 20,-

Vol. 63: J. Kohlas, Monte Carlo Simulation im Operations Research. VI, 162 Seiten. 4°. 1972. DM 16,-

Vol. 64: P. Gessner u. K. Spremann, Optimierung in Funktionenräumen. IV, 120 Seiten. 4°. 1972. DM 16,-

Vol. 65: W. Everling, Exercises in Computer Systems Analysis. VIII, 184 pages. 4°. 1972. DM 18,-

Vol. 66: F. Bauer, P. Garabedian and D. Korn, Supercritical Wing Sections. V, 211 pages. 4°. 1972. DM 20,-

Vol. 67: I. V. Girsanov, Lectures on Mathematical Theory of Extremum Problems. V, 136 pages. 4°. 1972. DM 16,-

Vol. 68: J. Loeckx, Computability and Decidability. An Introduction for Students of Computer Science. VI, 76 pages. 4°. 1972. DM 16,-

Vol. 69: S. Ashour, Sequencing Theory. V, 133 pages. 4°. 1972.
DM 16,–

Vol. 70: J. P. Brown, The Economic Effects of Floods. Investigations
of a Stochastic Model of Rational Investment Behavior in the Face
of Floods. V, 87 pages. 4°. 1972. DM 16,–

Vol. 71: R. Henn und O. Opitz, Konsum- und Produktionstheorie II.
V, 134 Seiten. 4°. 1972. DM 16,–

Vol. 72: T. P. Bagchi and J. G. C. Templeton, Numerical Methods in
Markov Chains and Bulk Queues. XI, 89 pages. 4°. 1972. DM 16,–

Vol. 73: H. Kiendl, Suboptimale Regler mit abschnittweise linearer
Struktur. VI, 146 Seiten. 4°. 1972. DM 16,–

Vol. 74: F. Pokropp, Aggregation von Produktionsfunktionen. VI, 107
Seiten. 4°. 1972. DM 16,–

Vol. 75: GI-Gesellschaft für Informatik e. V. Bericht Nr. 3. 1. Fach-
tagung über Programmiersprachen · München, 9–11. März 1971.
Herausgegeben im Auftrag der Gesellschaft für Informatik von H.
Langmaack und M. Paul. VII, 280 Seiten. 4°. 1972. DM 24,–

Vol. 76: G. Fandel, Optimale Entscheidung bei mehrfacher Zielset-
zung. 121 Seiten. 4°. 1972. DM 16,–

Vol. 77: A. Auslender, Problemes de Minimax via l'Analyse Convexe
et les Inégalités Variationnelles: Théorie et Algorithmes. VII, 132
pages. 4°. 1972. DM 16,–

Vol. 78 : GI-Gesellschaft für Informatik e.V. 2. Jahrestagung, Karlsruhe,
2.–4. Oktober 1972. Herausgegeben im Auftrag der Gesellschaft für
Informatik von P. Deussen. XI, 576 Seiten. 4°. 1973. DM 36,–

Vol. 79 : A. Berman, Cones, Matrices and Mathematical Programming.
V, 96 pages. 4°. 1973. DM 16,–

Vol. 80: International Seminar on Trends in Mathematical Modelling,
Venice, 13–18 December 1971. Edited by N. Hawkes. VI, 288 pages.
4°. 1973. DM 24,–

Vol. 81: Advanced Course on Software Engineering. Edited by
F. L. Bauer. XII, 545 pages. 4°. 1973. DM 32,–

Vol. 82: R. Saeks, Resolution Space, Operators and Systems. X, 267
pages. 4°. 1973. DM 22,–

Vol. 83: NTG/GI-Gesellschaft für Informatik, Nachrichtentechnische
Gesellschaft. Fachtagung „Cognitive Verfahren und Systeme", Ham-
burg, 11.–13. April 1973. Herausgegeben im Auftrag der NTG/GI von
Th. Einsele, W. Giloi und H.-H. Nagel. VIII, 373 Seiten. 4°. 1973.
DM 28,–

Vol. 84: A. V. Balakrishnan, Stochastic Differential Systems I. Filtering
and Control, A Function Space Approach. V, 252 pages. 4°. 1973.
DM 22,–

Vol. 85: T. Page, Economics of Involuntary Transfers: A Unified
Approach to Pollution and Congestion Externalities. XI, 159 pages.
4°. 1973. DM 18,–

Vol. 87: G. F. Newell, Approximate Stochastic Behavior of n-Server
Service Systems with Large n. VIII, 118 pages. 4°. 1973. DM 16,–